NATEF Standards Job Sheets

Automatic Transmissions and Transaxles (A2)

Second Edition

Jack Erjavec

THOMSON

DELMAR LEARNING

Australia   Canada   Mexico   Singapore   Spain   United Kingdom   United States

THOMSON

DELMAR LEARNING

NATEF Standards Job Sheets

Automatic Transmissions and Transaxles A2
Second Edition

Jack Erjavec

Vice President, Technology and Trades SBU:
Alar Elken

Editorial Editor:
Sandy Clark

Senior Acquisitions Editor:
David Boelio

Development Editor:
Matthew Thouin

Marketing Director:
David Garza

Channel Manager:
William Lawrensen

Marketing Coordinator:
Mark Pierro

Production Director:
Mary Ellen Black

Production Editor:
Toni Hansen

Art/Design Specialist:
Cheri Plasse

Technology Project Manager:
Kevin Smith

Editorial Assistant:
Andrea Domkowski

ISBN: 1-4180-2075-3

NOTICE TO THE READER

CONTENTS

PREFACE

The automotive service industry continues to change with the technological changes made by automobile and tool and equipment manufacturers. Today's automotive technician must have a thorough knowledge of automotive systems and components, good computer skills, exceptional communication skills, good reasoning, the ability to read and follow instructions, and above average mechanical aptitude and manual dexterity.

This new edition, like the last, was designed to give students a chance to develop the same skills and gain the same knowledge that today's successful technician has. This edition also reflects the changes in the guidelines established by the National Automotive Technicians Education Foundation (NATEF), as of July 2005.

The purpose of NATEF is to evaluate technician training programs against standards developed by the automotive industry and recommend qualifying programs for certification (accreditation) by ASE (National Institute for Automotive Service Excellence). Programs can earn ASE certification upon the recommendation of NATEF. NATEF's national standards reflect the skills that students must master. ASE certification through NATEF evaluation ensures that certified training programs meet or exceed industry-recognized, uniform standards of excellence.

At the expense of much time and many minds, NATEF has assembled a list of basic tasks for each of their certification areas. These tasks identify the basic skills and knowledge levels that competent technicians have. The tasks also identify what is required for a student to start a successful career as a technician.

Most of the content in this book are job sheets. These job sheets relate to the tasks specified by NATEF. The main considerations during the creation of these job sheets were student learning and program certification by NATEF.

Students are guided through standard industry accepted procedures. While they are progressing, they are asked to report their findings as well as offer their thoughts on the steps they have just completed. The questions asked of the students are thought provoking and require students to apply what they know to what they observe.

The job sheets were also designed to be generic. That is, whenever possible, the tasks can be performed on any vehicle from any manufacturer. Also, completion of the sheets does not require the use of specific brands of tools and equipment; rather students use what is available. In addition, the job sheets can be used as a supplement to any good textbook.

Also included are description and basic use of the tools and equipment listed in NATEF's standards. The standards recognize that not all programs have the same needs, nor do all programs teach all of the NATEF tasks. Therefore, the basic philosophy for the tools and equipment requirement is that the training should be as thorough as possible with the tools and equipment necessary for those tasks.

Theory instruction and hands-on experience of the basic tasks provide initial training for employment in automotive service or further training in any or all of the specialty areas. Competency in the tasks indicates to employers that you are skilled in that area. You need to know the appropriate theory, safety, and support information for each required task. This should include identification and use of the required tools and testing and measurement equipment required for the tasks, the use of current reference and training materials, the proper way to write work orders and warranty reports, and the storage, handling, and use of Hazardous Materials as required by the 'Right to Know Law', and federal, state, and local governments.

Words to the Instructor: I suggest you grade these job sheets based on completion and reasoning. Make sure the students answer all questions. Then look at their reasoning to see if the task was actually completed and to get a feel for their understanding of the topic. It will be easy for students to copy others' measurements and findings, but each student should have their own base of understanding and that will be reflected in their explanations.

Words to the Student: While completing the job sheets, you have a chance to develop the skills you need to be successful. When asked for your thoughts or opinions, think about what you observed. Think about what could have caused those results or conditions. You are not being asked to give accurate explanations for everything you do or observe. You are only asked to think. Thinking leads to understanding. Good technicians are good because they have a basic understanding of what they are doing and of why they are doing it.

Jack Erjavec

AUTOMATIC TRANSMISSIONS AND TRANSAXLES

To prepare you to learn what you should learn from completing the job sheets, some basics must be covered. This discussion begins with an overview of automatic transmissions. Emphasis is placed on what they do and how they work, including the major components and designs of automatic transmissions and their role in the efficient operation of automatic transmissions of all designs.

Preparing to work on an automobile would not be complete without addressing certain safety issues. This discussion covers what you should and should not do while working on automatic transmissions, including the proper ways to deal with hazardous and toxic materials.

NATEF's task list for automatic transmissions and transaxles certification is given, along with definitions of some of the terms used to describe the tasks. This list gives you a good look at what the experts say you need to know before you can be considered competent to work on automatic transmissions.

After the task list are descriptions of the various tools and types of equipment you need to be familiar with. These are the tools you will use to complete the job sheets. They are also the tools NATEF has identified as being necessary for servicing automatic transmissions.

After the tool discussion is a cross-reference guide that shows which NATEF tasks are related to specific job sheets. In most cases there is a single job sheet for each task. Some tasks are part of a procedure, in which case one job sheet may cover two or more tasks. The remainder of the book contains the job sheets.

BASIC AUTOMATIC TRANSMISSION THEORY

Many rear-wheel-drive (RWD) and four-wheel-drive (4WD) vehicles are equipped with automatic transmissions. Automatic transaxles, which combine an automatic transmission and final drive assembly in a single unit, are used on front-wheel-drive (FWD) and some all-wheel-drive (AWD) and RWD vehicles.

An automatic transmission or transaxle selects gear ratios according to engine speed, power train load, vehicle speed, and other operating factors. The most widely used automatic transmissions and transaxles are four-speed units. Three- and five-speed transmissions are also used. All new-model transmissions have computer-controlled torque converter clutches and shifting. Based on input data supplied by electronic sensors and switches, the computer sets the torque converter's operating mode, controls the transmission's shifting sequence, and in some cases regulates transmission fluid pressure.

Torque Converter

Automatic transmissions use a torque converter (Figure 1) to transfer engine torque from the engine to the transmission. The torque converter operates through hydraulic force provided by fluid. The torque converter automatically engages and disengages power from the engine to the transmission in relation to engine rpm. With the engine running at the correct idle speed, there is not

Figure 1   A torque converter and automatic transmission.

enough fluid flow for power transfer through the torque converter. As engine speed increases, the added fluid flow creates sufficient force to transmit engine power through the torque-converter assembly to the transmission.

The torque converter, located between the engine and transmission, is a sealed, doughnut-shaped unit that is always filled with automatic transmission fluid. A special flex plate is used to mount the torque converter to the crankshaft. The purpose of the flex plate is to transfer crankshaft rotation to the shell of the torque converter assembly. The flex plate also carries the starter motor ring gear. A flywheel is not required because the mass of the torque converter and flex disc acts like a flywheel to smooth out the engine's intermittent power strokes.

A standard torque converter consists of three elements: the impeller, the stator assembly, and the turbine. The impeller assembly is the input member; it receives power from the engine. The turbine is the output member; it is splined to the transmission's input shaft. The stator assembly is the reaction member or torque multiplier. The stator is supported on a one-way clutch, which operates as an overrunning clutch and permits the stator to rotate freely in one direction and lock up in the opposite direction.

The impeller forms one internal section of the torque converter shell. The impeller has numerous curved blades that rotate as a unit with the shell. It turns at engine speed, acting like a pump to start the transmission oil circulating within the torque converter shell.

The impeller is positioned with its back facing the transmission housing, whereas the turbine is positioned with its back to the engine. The curved blades of the turbine face the impeller

assembly. The turbine blades have a greater curve than the impeller blades, which helps eliminate oil turbulence between the turbine and impeller blades that would slow impeller speed and reduce the converter's efficiency.

The stator is located between the impeller and turbine. It redirects the oil flow from the turbine back into the impeller in the direction of impeller rotation with minimal loss of speed and force. The side of the stator blade with the inward curve is the concave side. The side with an outward curve is the convex side.

At the rear of the torque converter shell is a hollow hub with notches or flats at one end, set 180 degrees apart. This hub is called the pump drive hub. The notches or flats drive the transmission pump assembly. At the front of the transmission, within the pump housing, is a pump bushing that supports the pump drive hub and provides rear support for the torque converter assembly. Some other transaxles have a separate shaft to drive the pump.

As the impeller rotates, centrifugal force throws the oil outward and upward due to the curved shape of the impeller housing. The faster the impeller rotates, the greater the centrifugal force becomes. Oil thrown outward and upward from the impeller strikes the curved vanes of the turbine, causing the turbine to rotate. (There is no direct mechanical link between the impeller and the turbine.) Oil leaving the turbine is directed out of the torque converter to an external oil cooler and then to the transmission's oil sump or pan.

With the transmission in gear and the engine at idle, the vehicle can be held stationary by applying the brakes. At idle, engine speed is slow. Since the impeller is driven by engine speed, it turns slowly, creating little centrifugal force within the torque converter. Therefore, little or no power is transferred to the transmission.

When the throttle is opened, engine speed, impeller speed, and the amount of centrifugal force generated in the torque converter all increase dramatically. Oil is then directed against the turbine blades, which transfer power to the turbine shaft and transmission.

As the speed of the turbine approaches the speed of the impeller, the coupling point is reached and the turbine and the impeller run at essentially the same speed. They cannot run at exactly the same speed due to slippage between them. The only way they can turn at exactly the same speed is by using a torque converter clutch to mechanically tie them together.

The stator redirects the oil leaving the turbine back to the impeller, which helps the impeller rotate more efficiently. Torque converter multiplication can only occur when the impeller is rotating faster than the turbine.

A lockup torque converter eliminates the 10 percent slip that takes place between the impeller and turbine at the coupling stage of operation. There are many types of lockup torque converters. The lockup piston clutch is the type installed in most automatic transmissions. There are also fully mechanical lockup converters, centrifugal lockup converters, and converters that use a viscous coupling.

Planetary Gears

Nearly all automatic transmissions rely on planetary gear sets to transfer power and multiply engine torque to the drive axles. Compound gear sets combine two simple planetary gear sets so load can be spread over a greater number of teeth for strength and also to obtain the largest number of gear ratios possible in a compact area.

A simple planetary gear set consists of three parts: a sun gear, a carrier with planetary pinions mounted to it, and an internally toothed ring gear or annulus. The sun gear is located in the center of the assembly and meshes with the teeth of the planetary pinion gears. Planetary pinion gears are small gears fitted into a framework called the planetary carrier. The planetary carrier includes a shaft for each of the planetary pinion gears to rotate on. The planetary pinions surround the sun gear's center axis, and the ring gear surrounds them. The ring gear acts like a band to hold the entire gear set together and provide great strength to the unit.

A limited number of gear ratios are available from a single planetary gear set. To increase the number of available gear ratios, gear sets can be added. The typical automatic transmission with three or four forward speeds has at least two planetary gear sets. There are two common designs of compound gear sets: the Simpson gear set, in which two planetary gear sets share a common sun gear, and the Ravigneaux gear set, which has two sun gears, two sets of planetary gears, and a common ring gear. Some transmissions are fitted

with an additional single planetary gear set, which is used to provide an "add-on" overdrive gear.

Rather than relying on the use of a compound gear set, some automatic transmissions use two simple planetary units in series. In this type of arrangement, gear set members are not shared; instead, the holding devices are used to lock different members of the planetary units together.

A few automatic transaxles are based on the design of a manual transmission, that is, they use constant-mesh helical and square-cut gears.

Another unconventional transmission design, the continuously variable transmission (CVT), has no fixed forward speeds. The gear ratio varies with engine speed and temperature. Rather than gears, these transmissions use belts and pulleys. These transmissions also do not have a torque converter; instead, they use a manual transmission–type flywheel with a start clutch. One pulley is the driven member and the other is the drive. Each pulley has a movable face and a fixed face. When the movable face moves, the effective diameter of the pulley changes. The change in effective diameter changes the effective pulley (gear) ratio. A steel belt links the driven and drive pulleys.

Planetary Gear Controls

Certain parts of the planetary gear train must be held while others must be driven to provide the needed torque multiplication and direction for vehicle operation. Bands, brakes, and clutches are the typical planetary gear controls used today.

A band is a braking assembly positioned around a stationary or rotating drum. The band brings a drum to a stop by wrapping itself around the drum and holding it. A servo assembly hydraulically applies the band. Connected to the drum is a member of the planetary gear train. The purpose of a band is to hold a member of the planetary gear set by holding the drum and connecting planetary gear member stationary. The servo assembly converts hydraulic pressure into a mechanical force that applies a band to hold a drum stationary.

In an automatic transmission operation, both sprag and roller overrunning clutches are used to hold members of the planetary gear set. These clutches operate mechanically. An overrunning clutch allows rotation in only one direction and operates at all times. One-way overrunning clutches can be either roller-type or sprag-type clutches.

A multiple–friction disc assembly uses a series of friction discs to transmit torque or apply braking force. The friction discs have internal teeth that are sized and shaped to mesh with splines on the clutch assembly hub. In turn, this hub is connected to a member of the planetary gear set that receives the desired braking or transfer force when the clutch is applied or released.

Multiple-disc clutches are enclosed in a large drum-shaped housing that holds the other clutch components: cylinder, hub, piston, piston return springs, seals, pressure plate, clutch plates, friction plates, and snap rings.

The friction discs are sandwiched between the clutch plates and pressure plate. Friction discs are steel core plates with friction material bonded to either side. The pressure plate has tabs around the outside diameter to mate with the channels in the clutch drum. It is held in place by a large snap ring. Upon engagement, the clutch piston forces the clutch pack against the fixed pressure plate.

Bearings, Bushings, and Thrust Washers

When a component slides over or rotates around another part, the surfaces that contact each other are called bearing surfaces. A gear rotating on a fixed shaft can have more than one bearing surface if it is supported and held in place by the shaft in a radial direction. In addition, the gear tends to move along the shaft in an axial direction as it rotates and is therefore held in place by some other components. The surfaces between the sides of the gear and the other parts are bearing surfaces.

A bearing is placed between two bearing surfaces to reduce friction and wear. In automatic transmissions, sliding bearings are used where one or more of the following conditions prevail: low rotating speeds, very large bearing surfaces compared to the surfaces present, and low use. Rolling bearings are used in high-speed applications, high load with relatively small bearing surfaces, and high use.

Transmissions use sliding bearings made of relatively soft bronze alloy. Many are made from steel with the bearing surface bonded or fused to the steel. Bearings that take radial loads are called bushings and those that take axial loads are called thrust washers.

Since bushings are made of a soft metal, they act like a bearing and support many of the transmission's rotating parts. They are also used to precisely guide the movement of various valves in the transmission's valve body. Bushings can also be used to control fluid flow; some bushings restrict the flow from one part to another, and others are made to direct fluid flow to a particular point or part in the transmission.

Thrust washers are made in various thicknesses. They may have one or more tangs or slots on the inside or outside circumference that mate with the shaft bore to keep them from turning. Some thrust washers are made of nylon or Teflon, which are used when the load is low. Others are fitted with rollers to reduce friction and wear.

Thrust washers normally control free axial movement or endplay. Since some endplay is necessary in all transmissions because of heat expansion, proper endplay is often accomplished through selective thrust washers. These thrust washers are inserted between various parts of the transmission. Thrust washers work by filling the gap between two objects and become the primary wear item because they are made of softer materials than the parts they protect.

Torrington bearings are thrust washers fitted with roller bearings. These thrust bearings are primarily used to limit endplay but also to reduce the friction between two rotating parts. Most often Torrington bearings are used in combination with flat thrust washers to control endplay of a shaft or the gap between a gear and its drum.

Snap Rings

Many different sizes and types of snap rings are used in today's transmissions. External and internal snap rings are used as retaining devices throughout the transmission. Internal snap rings are used to hold servo assemblies and clutch assemblies together. In fact, snap rings are also available in several thicknesses and may be used to adjust the clearance in multiple-disc clutches. Some clutch packs use waved snap rings to provide for smoother clutch applications. External snap rings are used to hold gear and clutch assemblies to their shafts.

Gaskets and Seals

The gaskets and seals of an automatic transmission help to keep the fluid in the transmission and prevent it from leaking out of the various hydraulic circuits. Different types of seals are used in automatic transmissions; they can be made of rubber, metal, or Teflon. Transmission gaskets are made of rubber, cork, paper, synthetic materials, or plastic.

Gaskets are used to seal two parts together or to provide a passage for fluid flow from one part of the transmission to another. Soft gaskets are used when the sealing surfaces are irregular or in places where the surface may distort when the component is tightened into place. A typical location for a soft gasket is the oil pan gasket that seals the oil pan to the transmission case. Oil pan gaskets are typically a composition-type gasket made with rubber and cork. However, some late-model transmissions use a room temperature vulcanizing (RTV) sealant instead of a gasket to seal the oil pan.

As valves and transmission shafts move within the transmission, it is essential that the fluid and pressure be contained within its bore. Any leakage would decrease the pressure and result in poor transmission operation. Seals are used to prevent leakage around valves, shafts, and other moving parts.

O-rings are round seals with a circular cross section. Normally an O-ring is installed in a groove cut into the inside diameter of one of the parts to be sealed. When the other part is inserted into the bore and through the O-ring, the O-ring is compressed between the inner part and the groove. This pressure distorts the O-ring and forms a tight seal between the two parts.

Lip seals are used to seal parts that have axial or rotational movement. They are round so that they fit around shafts, but the entire seal does not serve as a seal; instead, the sealing part is a flexible lip. The flexible lip is normally made of synthetic rubber and shaped so that it is flexed when installed in order to apply pressure at the sharp edge of the lip. Lip seals are used around input and output shafts to keep fluid in the housing and dirt out. Some seals are double-lipped.

A square-cut seal can withstand more axial movement than an O-ring can. Square-cut seals are also round seals but have a rectangular or square cross section. They are designed this way to prevent the seal from rolling in its groove when there is a large amount of axial movement. Added sealing comes from the distortion of the seal during axial movement. As the shaft inside the seal

moves, the outer edge of the seal moves more than the inner edge, thereby causing the diameter of the sealing edge to increase, which creates a tighter seal.

Some parts of the transmission do not require a positive seal; some leakage is acceptable. These components are sealed with ring seals that fit into a groove on a shaft. The outside diameter of the ring seals slide against the walls of the bore that the shaft is inserted into. Most ring seals in a transmission are placed near pressurized fluid outlets on rotating shafts to help retain pressure. Three types of metal seals are used in automatic transmissions: butt-end seals, open-end seals, and hook-end seals.

Some transmissions use Teflon seals instead of metal seals. Teflon provides for a softer sealing surface, which results in less wear on the surface that it rides on and therefore a longer-lasting seal. Teflon seals are similar in appearance to metal seals except for the hook-end type. The ends of locking-end Teflon seals are cut at an angle; locking-end seals are often called scarf-cut seals.

Many late-model transmissions are equipped with solid one-piece Teflon seals. Although the one-piece seal requires some special tools for installation, it makes a near-positive seal. These Teflon rings seal much better than other metal sealing rings.

Final Drive Assemblies

The last set of gears in the drive train is the final drive. In most RWD cars, the final drive is located in the rear axle housing. On most FWD cars, the final drive is located within the transaxle. Some FWD cars with longitudinally mounted engines locate the differential and final drive in a separate case that bolts to the transmission.

A transaxle's final drive gears provide a way to transmit the transmission's output to the differential section of the transaxle. Four common configurations are used as the final drives on FWD vehicles: helical gear, planetary gear, hypoid gear, and chain drive. The helical, planetary, and chain final drive arrangements are found with transversely mounted engines. Hypoid final drive gear assemblies are normally found in vehicles with a longitudinally placed engine. The hypoid assembly is basically the same unit as that used on RWD vehicles and is mounted directly to the transmission.

Some transaxles route output torque through two helical-cut gears to a transfer shaft. A helical-cut pinion gear attached to the opposite end of the transfer shaft drives the differential ring gear and carrier.

Other transaxles use a simple planetary gear set for the final drive. In operation, the transmission's output drives the sun gear that, in turn, drives the planetary pinion gears. The pinion gears walk around the inside of the stationary ring gear. The rotating planetary pinion gears drive the planetary carrier and differential case. This combination provides maximum torque multiplication from a simple planetary gear set.

Chain-drive final drive assemblies use a multiple-link chain to connect a drive sprocket, which is connected to the transmission's output shaft, to a driven sprocket, which is connected to the differential case. This design allows for remote positioning of the differential within the transaxle housing. Final drive gear ratios are determined by the size of the driven sprocket compared to the drive sprocket.

Hydraulic System

An automatic transmission uses fluid pressure to control the action of the planetary gear sets. This fluid pressure is regulated and directed to change gears automatically through the use of various pressure regulators and control valves.

The automatic transmission reservoir is the transmission oil pan. Transmission fluid is drawn from the pan and returned to it. The pressure source in the system is the oil pump. The valve body contains control valves to regulate or restrict the pressure and flow of fluid within the transmission. Output devices are the servos or clutches operated by hydraulic pressure.

The transmission pump is driven by the torque converter shell at engine speed. The purpose of the pump is to create fluid flow in the system. Pump pressure is a variable pressure, depending on engine speed from idle to full throttle. The pressure in a transmission's hydraulic system is the result of constant fluid flow from the pump and the valves, pistons, seals, and bushings in the transmission.

Basic Hydraulic Theory

An automatic transmission is a complex hydraulic circuit. Hydraulics is the study of liquids in motion. A fluid is something that does not have a definite shape, therefore liquids and gases are fluids. A characteristic of all fluids is that they will conform to the shape of their container. A liquid also will typically not compress, regardless of the pressure on it. Therefore, liquids are considered non-compressible fluids.

Liquids will, however, predictably respond to pressures exerted on them. Their reaction to pressure is the basis of all hydraulic applications. This fact allows hydraulics to do work.

Over 300 years ago a French scientist, Blaise Pascal, determined that if you had a liquid-filled container with only one opening and applied force to the liquid through that opening, the force would be evenly distributed throughout the liquid. This explains how pressurized liquid is used to operate and control an automatic transmission. A pump pressurizes the transmission fluid and the fluid is delivered to the various apply devices. When the pressure increases enough to activate the apply device, it holds a gear set member until the pressure decreases. The valve body is responsible for the distribution of the pressurized fluid to the appropriate reaction member according to the operating conditions of the vehicle.

Pascal also found he could increase the force available to do work, simply by moving fluid under pressure from a cylinder to a larger one. He determined that force applied to liquid creates pressure or the transmission of force through the liquid. His experiments revealed two important aspects of a liquid when it is confined and put under pressure: the pressure applied to it is transmitted equally in all directions and this pressure acts with equal force at every point in the container.

When a pressure is applied to a confined liquid, the pressure of the liquid is the same everywhere within the hydraulic system. If the hydraulic pump provides 100 psi, there will be 100 pounds of pressure on every square inch of the system. If the system has a piston with an area of 30 square inches, each square inch receives 100 pounds of pressure. This means there will be 3,000 pounds of force applied to that piston. According to Pascal's law, force is equal to the pressure multiplied by the area of piston ($F = P \times A$). The use of the larger piston gives the system a mechanical advantage or increase in the force available to do work.

By changing the size of the pistons in a hydraulic system, force is multiplied, and as a result, low amounts of force are needed to move heavy objects.

Transmission Fluid

The automatic transmission fluid (ATF) circulating through the transmission and torque converter and over the parts of the transmission cools the transmission. The heated fluid moves to the transmission fluid cooler where the heat is removed. As the fluid lubricates and cools the transmission, it also cleans the parts. The dirt is carried by the fluid to a filter where the dirt is removed.

Another critical job of ATF is its role in shifting gears. ATF moves under pressure throughout the transmission and causes various valves to move. The pressure of the ATF changes with changes in engine speed and load.

ATF is also used to operate the various apply devices (clutches and brakes) in the transmission. At the appropriate time, a switching valve opens and sends pressurized fluid to the apply device which engages or disengages a gear.

Valve Body

The purpose of the valve body is to respond to the speed of the engine and the load on it and the drive train, as well as the driver's intentions. The valve body is an aluminum or iron casting with many precisely machined holes and passages that accommodate fluid flow and various valves. Internally, the valve body has many fluid passages.

The purpose of a valve is to start, stop, or direct and regulate fluid flow. In most valve bodies, three types of valves are used: check ball, poppet, and, most commonly, the spool.

The *check ball valve* is a ball that operates on a seat located on the valve body. The check ball operates by having a fluid pressure or manually operated linkage force it against the ball seat to block fluid flow. Pressure on the opposite side unseats the check ball. Check balls and poppet valves can be designed to be normally open, which allows free flow of fluid pressure, or normally closed, which blocks fluid pressure flow. Other applications of the check ball have two seats to check and direct fluid flow from two directions, being seated and unseated by pressures from either source.

Check ball valves can also be used as pressure relief valves to relieve excessive fluid pressure.

The check ball is held against its seat by spring tension that is stronger than the fluid pressure. When the fluid pressure overcomes the spring tension, the check ball is forced off its seat, which relieves excess pressure. As soon as the opposing fluid pressure is relieved, the spring tension forces the check ball back onto its seat.

A *poppet valve* can be a ball or a flat disc. In either case, the poppet valve acts to block fluid flow. Often the poppet valve has a stem to guide the valve's operation. The stem normally fits into a hole acting as a guide to the valve's opening and closing. Poppet valves tend to pop open and closed, hence their name.

The most commonly used valve in a valve body is the *spool valve*. The lands of a spool valve are large diameter surfaces that fit into a bore. There is a minimum of two lands per valve, and a stem connects the lands. Between the two lands is the valve's valley, which forms a fluid pressure chamber between the lands and valve body bore. Fluid flow can be directed into other passages, depending on the spool valve and valve body design.

The land rides on a very thin film of fluid in the valve body bore. The land must be treated very carefully because any damage, even a small score or scratch, can impair smooth valve operation. As the spool valve moves, the land covers (closes) or uncovers (opens) ports in the valve body. When a spool valve cycles back and forth between the open and the exhaust positions, it is called a regulating valve.

Pressures

Line pressure is also referred to as mainline pressure. Line pressure is the hydraulic pressure that operates apply devices and is the source for all other pressures in the transmission. It is developed by the transmission's pump and is regulated by the pressure regulator.

Converter pressure transmits torque and keeps transmission fluid circulating into and out of the torque converter, which reduces the formation of air bubbles and aids cooling.

There are times in the automatic transmission's operation when fluid pressure must be increased above its baseline pressure. This increase is needed to hold bands and clutches more tightly and to raise the point at which shifting takes place. Increasing pressure above normal line pressures allows operational load flexibility, which is needed, for example, when towing trailers.

There are two methods used to monitor vehicle and engine load: vacuum modulation and throttle pressure. On some transmissions, a vacuum modulator is used to measure fluctuating engine vacuum to sense the load placed on the engine and drive train.

Many vehicles use throttle valve pressure to increase line pressure. Throttle valve pressure develops when line pressure passes through the throttle valve valley to become throttle pressure. Throttle pedal movement is carried through the throttle linkage to control the operation of the throttle valve.

When the pedal is depressed (opened), the throttle valve opens to produce throttle pressure, which is directed to the pressure regulator throttle plug. This pressure helps the pressure regulator valve spring hold the pressure regulator valve in position to close the exhaust port. This results in increased pressure.

When the pedal is released (closed), the throttle valve partially closes. This decreases throttle pressure at the throttle valve plug, resulting in a reduction of line pressure.

A relay valve controls the direction of line pressure flow. A relay valve is a spool-type valve with several lands and reaction areas. It is held in one position in the valve body bore by coil spring tension, auxiliary fluid pressure, or a mechanical force. Auxiliary fluid pressure or mechanical force can oppose coil spring tension to move the relay valve to a new position. In the new position, the valley of the relay valve aligns with interconnecting ports. Fluid pressure flows from an inlet port across the relay valve valley to the outlet port. When a relay valve is in one position, it blocks fluid flow. When it is moved to another position, fluid is directed through the relay valve to the outlet port.

One type of relay valve is a shift valve. A shift valve usually operates in one of two positions—either downshifted or upshifted. In operation, throttle pressure, which is high, is acting on one reaction area of a shift valve. Throttle pressure and light coil spring tension act to hold the valve in the downshifted position. In this position, fluid pressure is blocked from flowing to one of the planetary gear controls.

As vehicle speed increases, the governor develops a pressure, called governor pressure. Governor pressure is directed to the opposite reaction area of the shift valve. With increasing vehicle speed, governor pressure increases to overcome coil spring tension and throttle pressure. Governor pressure moves the shift valve to the upshifted position. In this position the inlet and outlet ports are in the same valley area of the shift valve. Fluid pressure flows across the shift valve valley, out of the outlet port, and through the connecting worm tracks to engage the planetary controls for the next higher gear.

If the driver pushes down on the throttle pedal, throttle pressure increases. When throttle pressure is higher than governor pressure, the high throttle pressure and coil spring tension force the shift valve to move to the downshifted position against governor pressure. The transmission automatically downshifts to the next lower gear. A kickdown valve is often used to generate additional kickdown pressure to ensure that the downshift is made quickly and positively.

The manual valve is a spool valve operated manually by the driver and gear selector linkage. When the driver selects a gear position, the gear selector linkage positions the manual valve in the valve body. The manual valve directs line pressure to the correct combination of circuits, which produces the range desired for the driver's requirements.

Governor

The governor assembly senses vehicle road speed and sends a fluid pressure signal to the transmission valve body to either upshift or permit the transmission to downshift. The governor assembly is located on the transmission output shaft and is influenced by increases and decreases in output shaft speed. As vehicle speed increases, the transmission output shaft speed increases, resulting in more governor pressure to automatically upshift the transmission.

Shift Feel

All transmissions are designed to change gears at the correct time, according to engine speed, load, and driver intent. However, transmissions are also designed to provide for a positive change of gear ratios without jarring the driver or passengers. If a band or clutch is applied too quickly, a harsh shift will occur.

Shift feel is controlled by the pressure at which each reaction member is applied or released, the rate at which each is pressurized or exhausted, and the relative timing of the application and release of the members.

To improve shift feel during gear changes, a band is often released while a multiple-disc clutch is being applied. The timing of these two actions must be just right or both components will be released or applied at the same time, which would cause engine flare-up or driveline shudder.

Accumulators

Shift quality with a band or clutch depends on how quickly the apply device is engaged by hydraulic pressure and the amount of pressure exerted on the piston. Some apply circuits utilize an accumulator to slow down application rates without decreasing the holding force of the apply device.

An accumulator is similar to a servo in that it consists of a piston and cylinder. It works like a shock absorber and cushions the application of servos and clutches. An accumulator cushions sudden increases in hydraulic pressure by temporarily diverting some of the apply fluid into a parallel circuit or chamber. This allows the pressure to increase gradually and provides for smooth engagement of a band or clutch.

Electronic Control of Torque Converter Clutch Engagement

In many vehicles, automatic transmissions and transaxles are electronically controlled. Shifting and torque converter clutch engagement are regulated by a computer with programmed logic and in response to input sensors and switches. The electronic system controls these two operations by means of solenoid-operated valves. With electronic control, information about the engine, fuel, ignition, vacuum, and operating temperature is used to ensure that shifting and clutch engagement take place at exactly the right time.

A computer is used to control the converter clutch solenoid. The computer turns on the converter clutch solenoid to move a valve that allows fluid pressure to engage the converter clutch.

When the computer de-energizes the converter clutch solenoid, the converter clutch disengages.

Electronically Controlled Shifting

Electronically controlled transmissions function in the same way as hydraulically based transmissions, except that a computer determines their shift points. The computer uses inputs from several different sensors and matches this information to a predetermined schedule.

The computer determines the time and condition of gear changes according to the input signals it receives and the shift schedules it has stored in its memory. Shift schedule logic chooses the proper shift schedule for the current conditions of the transmission. It uses the shift schedule to select the appropriate gear and then determines the correct shift schedule or pattern that should be followed.

Electronically controlled transmissions typically don't have governors or throttle pressure devices. They do, however, rely on pressure differentials at the sides of a shift valve to hold or change a gear. The pressure differential is caused by the action of shift solenoids, which allow for changes in pressure on the side of a shift valve. The computer controls these solenoids. The solenoids do not directly control the transmission's clutches and bands. These are engaged or disengaged hydraulically. The solenoids simply control the fluid pressures in the transmission and do not perform a mechanical function.

Most electronically controlled systems are complete computer systems. There is a central processing unit, inputs, and outputs. Often the central processing unit is a separate computer designated for transmission control. This computer is often called the transmission control module (TCM). Other transmission control systems use the powertrain CM (PCM) or the body CM (BCM) to control shifting. When transmission control is not handled by the PCM, the controlling unit communicates with the PCM.

The inputs include transmission operation monitors plus some of the sensors used by the PCM. Input sensors, such as the throttle position (TP) sensor, supply information for many different systems, and the control modules share this information. The system outputs are solenoids. The transmission control systems used by various vehicle manufacturers differ mostly in the number and type of solenoids used.

Many late-model transmissions have systems that allow the TCM to change transmission behavior in response to operating conditions and to the habits of the driver. The system monitors the condition of the engine and compensates for any changes in the engine's or transmission's performance. It also monitors and memorizes the typical driving style of the driver and the operating conditions of the vehicle. With this information, the TCM adjusts the timing of shifts and converter lockup to provide good shifting at the appropriate time. Computer systems with this capability are said to have adaptive learning capabilities.

SAFETY

In an automotive repair shop, there is great potential for serious accidents simply because of the nature of the business and the equipment used. Through carelessness, the automotive repair industry can be one of the most dangerous occupations, but the chances of being injured while working on a car are close to nil if you learn to work safely and use common sense. Shop safety is the responsibility of everyone in the shop.

Personal Protection

Some procedures, such as grinding, result in tiny particles of metal and dust being thrown off at very high speeds. These metal and dirt particles can easily get into your eyes, causing scratches or cuts on your eyeball. Pressurized gases and liquids escaping a ruptured hose or hose fitting can spray a great distance. If these chemicals get into your eyes, they can cause blindness. Dirt and sharp bits of corroded metal can easily fall into your eyes while you are working under a vehicle.

Eye protection should be worn whenever you are exposed to these risks. To be safe, you should wear safety glasses whenever you are working in the shop. Some procedures may require that you wear other eye protection in addition to safety glasses. When cleaning parts with a pressurized spray, for example, you should wear a face shield. The face shield not only gives added protection to your eyes, it also protects the rest of your face.

If chemicals such as battery acid, fuel, or solvents get into your eyes, flush them continuously

with clean water. Have someone call a doctor, and get medical help immediately.

Your clothing should be well fitted and comfortable but made with strong material. Loose, baggy clothing can easily get caught in moving parts and machinery. Some technicians prefer to wear coveralls or shop coats to protect their personal clothing. Your work clothing should offer you some protection but should not restrict your movement.

Long hair and loose, hanging jewelry can create the same type of hazard as loose-fitting clothing—they can get caught in moving engine parts and machinery. If you have long hair, tie it back or tuck it under a cap.

Never wear rings, watches, bracelets, and neck chains. These can easily get caught in moving parts and cause serious injury.

Always wear leather or similar material shoes or boots with nonslip soles. Steel-tipped safety shoes can give added protection to your feet. Jogging or basketball shoes, street shoes, and sandals are inappropriate in the shop.

Good hand protection is often overlooked. A scrape, cut, or burn can limit your effectiveness at work for many days. A well-fitted pair of heavy work gloves should be worn during operations such as grinding and welding or when handling high-temperature components. Always wear approved rubber gloves when handling strong and dangerous caustic chemicals.

Many technicians wear thin, surgical-type latex gloves whenever they are working on vehicles. These offer little protection against cuts but do offer protection against disease and grease buildup under and around your fingernails. These gloves are comfortable and are quite inexpensive.

Accidents can be prevented simply by the way you act. Following are some guidelines for working in a shop. This list does not include everything you should or shouldn't do; it merely provides some things to think about.

- Never smoke while working on a vehicle or while working with any machine in the shop.

- Playing around is not fun when it sends someone to the hospital.

- To prevent serious burns, keep your skin away from hot metal parts, such as the radiator, exhaust manifold, tailpipe, catalytic converter, and muffler.

- Always disconnect electric engine cooling fans when working around the radiator. Many of these turn on without warning and can easily chop off a finger or hand. Make sure you reconnect the fan after you have completed your repairs.

- When working with a hydraulic press, make sure the pressure is applied in a safe manner. It is generally wise to stand to the side when operating the press.

- Properly store all parts and tools by putting them away in a place where people will not trip over them. This practice not only cuts down on injuries, it also reduces time wasted looking for a misplaced part or tool.

Work Area Safety

Your entire work area should be kept clean and dry. Any oil, coolant, or grease on the floor can make it slippery. To clean up oil, use commercial oil absorbent. Keep all water off the floor. Water is slippery on smooth floors, and electricity flows well through water. Aisles and walkways should be kept clean and wide enough to easily move through. Make sure the work areas around machines are large enough to operate the machine safely.

Gasoline is a highly flammable volatile liquid. Something that is *flammable* catches fire and burns easily. A *volatile* liquid is one that vaporizes very quickly. *Flammable volatile liquids* are potential firebombs. Always keep gasoline or diesel fuel in an approved safety can, and never use gasoline to clean your hands or tools.

Handle all solvents (or any liquids) with care to avoid spillage. Keep all solvent containers closed, except when pouring. Proper ventilation is very important in areas where volatile solvents and chemicals are used. Solvents and other combustible materials must be stored in approved and designated storage cabinets or rooms with adequate ventilation. Never light matches or smoke near flammable solvents and chemicals, including battery acids.

Oily rags should also be stored in an approved metal container. When oily, greasy, or paint-soaked rags are left lying about or are not stored properly, they can spontaneously combust. Spontaneous combustion refers to a fire that starts by itself, without a match.

Disconnecting the vehicle's battery before working on the electrical system or before welding can prevent fires caused by a vehicle's electrical system. To disconnect the battery, remove the negative or ground cable from the battery and position it away from the battery.

Know where all the shop's fire extinguishers are located. Fire extinguishers are clearly labeled as to their type and the types of fire they should be used on. Make sure you use the correct type of extinguisher for the type of fire you are dealing with. A multipurpose dry chemical fire extinguisher puts out ordinary combustibles, flammable liquids, and electrical fires. Never put water on a gasoline fire—water just spreads the fire. The proper fire extinguisher smothers the flames.

During a fire, never open doors or windows unless it is absolutely necessary; the extra draft only makes the fire worse. Make sure the fire department is contacted before or during your attempt to extinguish a fire.

Tool and Equipment Safety

Careless use of simple hand tools, such as wrenches, screwdrivers, and hammers, causes many shop accidents that could be prevented. Keep all hand tools free of grease and in good condition. Tools that slip can cause cuts and bruises. If a tool slips and falls into a moving part, it can fly out and cause serious injury.

Use the proper tool for the job. Make sure the tool is of professional quality. Using poorly made tools or the wrong tools can damage parts, the tool itself, or you. Never use broken or damaged tools.

Safety around power tools is very important. Serious injury can result from carelessness. Always wear safety glasses when using power tools. If the tool is electrically powered, make sure it is properly grounded. Before using it, check the wiring for cracks in the insulation, as well as for bare wires. Also, when using electrical power tools, never stand on a wet or damp floor. Never leave a running power tool unattended.

Tools that use compressed air are called pneumatic tools. Compressed air is used to inflate tires, apply paint, and drive tools. Compressed air can be dangerous when it is not used properly.

When using compressed air, wear safety glasses or a face shield, or both. Particles of dirt and pieces of metal blown by the high-pressure air can penetrate your skin or get into your eyes.

Before using a compressed air tool, check all hose connections. Always hold an air nozzle or air control device securely when starting or shutting off the compressed air. A loose nozzle can whip suddenly and cause serious injury. Never point an air nozzle at anyone. Never use compressed air to blow dirt from your clothes or hair. Never use compressed air to clean the floor or workbench.

Always be careful when raising a vehicle on a lift or a hoist. Adapters and hoist plates must be positioned correctly to prevent damage to the underbody of the vehicle. There are specific lift points that allow the weight of the vehicle to be evenly supported by the adapters or hoist plates. The correct lift points can be found in the vehicle's service manual. Before operating any lift or hoist, carefully read the operating manual and follow the operating instructions.

Once you feel the lift supports are properly positioned under the vehicle, raise the lift until the supports contact the vehicle. Then, check the supports to make sure they are in full contact with the vehicle. Shake the vehicle to make sure it is securely balanced on the lift, and then raise the lift to the desired working height. Before working under a car, make sure the lift's locking devices are engaged.

A vehicle can be raised off the ground by a hydraulic jack. The jack's lifting pad must be positioned under an area of the vehicle's frame or at one of the manufacturer's recommended lift points. Never place the pad under the floor pan or under steering and suspension components, which are easily damaged by the weight of the vehicle. Always position the jack so the wheels of the vehicle can roll as it is being raised.

Safety stands, also called jack stands, should be placed under a sturdy chassis member, such as the frame or axle housing, to support the vehicle after it has been raised by a jack. Once the safety stands are in position, the hydraulic pressure in the jack should be slowly released until the weight of the vehicle is on the stands. Never move under a vehicle when it is supported only by a hydraulic jack. Rest the vehicle on the safety stands before moving under the vehicle.

Heavy parts of the automobile, such as engines, are removed with chain hoists or cranes. Cranes often are called cherry pickers. To prevent serious injury, chain hoists and cranes must be properly attached to the parts being lifted. Always use bolts with enough strength to support the object being lifted. After you have attached the lifting chain or cable to the part that is being removed, have your instructor check it. Place the chain hoist or crane directly over the assembly, then attach the chain or cable to the hoist.

Parts cleaning is a necessary step in most repair procedures. Always wear the appropriate protection when using chemical, abrasive, and thermal cleaners.

Vehicle Operation

When a customer brings in a vehicle for service, certain driving rules should be followed to ensure your safety and the safety of those working around you. For example, before moving a car into the shop, buckle your safety belt. Make sure no one is nearby, the way is clear, and there are no tools or parts under the car before you start the engine.

Check the brakes before putting the vehicle in gear. Then drive slowly and carefully in and around the shop.

If the engine must be running while you are working on the car, block the wheels to prevent the car from moving. Put the transmission in park for automatic transmissions and set the parking (emergency) brake. Never stand directly in front of or behind a running vehicle.

Run the engine only in a well-ventilated area to avoid the danger of poisonous carbon monoxide (CO) in the engine exhaust. CO is an odorless but deadly gas. Most shops have an exhaust ventilation system; always use it. Connect the hose from the vehicle's tailpipe to the intake for the vent system. Make sure the vent system is turned on before running the engine. If the work area does not have an exhaust venting system, use a hose to direct the exhaust out of the building.

HAZARDOUS MATERIALS AND WASTES

A typical shop contains many potential health hazards for those working in it. These hazards can cause injury, sickness, health impairments, discomfort, and even death. Here is a short list of the different classes of hazards.

- Chemical hazards are caused by high concentrations of vapors, gases, or solids in the form of dust.

- Hazardous wastes are substances that result from a service.

■ Physical hazards include excessive noise, vibration, pressures, and temperatures.

■ Ergonomic hazards are conditions that impede normal or proper body position and motion.

Many government agencies are charged with ensuring safe work environments for all workers, including the Occupational Safety and Health Administration (OSHA), Mine Safety and Health Administration (MSHA), and National Institute for Occupational Safety and Health (NIOSH). These agencies, in addition to state and local governments, have instituted regulations that must be understood and followed. Everyone in a shop is responsible for adhering to these regulations.

An important part of a safe work environment is the employees' knowledge of potential hazards. Right-to-know laws concerning all chemicals protect every employee in the shop. The general intent of right-to-know laws is for employers to provide their employees with a safe working place as it relates to hazardous materials.

All employees must be trained about their rights under the legislation, the nature of the hazardous chemicals in their workplace, and the contents of the labels on the chemicals. All the information about each chemical must be posted on Material Safety Data Sheets (MSDS) and must be accessible. The manufacturer of the chemical must give these sheets to its customers on request. They detail the chemical composition and precautionary information for all products that can present a health or safety hazard.

Employees must become familiar with the general uses, protective equipment, accident and spill procedures, and any other information regarding the safe handling of the hazardous material. This training must be given to employees annually and provided to new employees as part of their job orientation.

All hazardous material must be properly labeled, indicating what health, fire, or reactivity hazard it poses and what protective equipment is necessary when handling each chemical. The manufacturer of the hazardous materials must provide all warnings and precautionary information, which must be read and understood by the user before use. A list of all hazardous materials used in the shop must be posted for employees to see.

Shops must maintain documentation on the hazardous chemicals in the workplace, proof of training programs, records of accidents and spill incidents, satisfaction of employee requests for specific chemical information via the MSDS, and a general right-to-know compliance procedure.

When handling any hazardous materials or hazardous waste, make sure you follow the required procedures for handling such material. Wear the proper safety equipment listed on the MSDS, including approved respirator equipment.

Some of the common hazardous materials that automotive technicians use are cleaning chemicals, fuels (gasoline and diesel), paints and thinners, battery electrolyte (acid), used engine oil, used transmission fluid, refrigerants, and engine coolant (antifreeze).

Many repair and service procedures generate what are known as hazardous wastes. Dirty solvents and cleaners are good examples of hazardous wastes. Something is classified as a hazardous waste if it is on the Environmental Protection Agency (EPA) list of known harmful materials or has one or more of the following characteristics.

■ *Ignitability.* A liquid with a flash point below 140 degrees F or a solid that can spontaneously ignite.

■ *Corrosivity.* A substance that dissolves metals and other materials or burns the skin.

■ *Reactivity.* Any material that reacts violently with water or other materials or releases cyanide gas, hydrogen sulfide gas, or similar gases when exposed to low-pH acid solutions, including material that generates toxic mists, fumes, vapors, and flammable gases.

■ *EP toxicity.* Materials that leach one or more of eight heavy metals in concentrations greater than 100 times primary drinking water standard concentrations.

Complete EPA lists of hazardous wastes can be found in the Code of Federal Regulations. It should be noted that no material is considered hazardous waste until the shop is finished using it and ready to dispose of it.

The following list describes the recommended procedure for dealing with some common hazardous wastes. Always follow these and any other mandated procedures.

Fluids Collect all used fluids, such as automatic transmission fluid (ATF), and store them with similar fluids in a container marked to identify the contents. Do not mix fluids, except as allowed by the recycler. Handle used fluids in the same way as engine oil.

Engine oil Recycle oil. Set up equipment such as a drip table or screen table with a used oil collection bucket to collect oils dripping from parts. Place drip pans underneath vehicles that are leaking fluids onto the storage area. Do not mix other wastes with used oil, except as allowed by your recycler. Used oil generated by a shop (and oil received from household "do-it-yourself" generators) may be burned on site in a commercial space heater. Used oil also may be burned for energy recovery. Contact state and local authorities to determine requirements and to obtain necessary permits.

Oil filters Drain for at least 24 hours, crush, and recycle used oil filters.

Batteries Recycle batteries by sending them to a reclaimer or back to the distributor. Keep shipping receipts to demonstrate that you have recycled. Store batteries in a watertight, acid resistant container. Inspect batteries for cracks and leaks when they come in. Treat a dropped battery as if it were cracked. Acid residue is hazardous because it is corrosive and may contain lead and other toxics. Neutralize spilled acid by using baking soda or lime, and dispose of as hazardous material.

Metal residue from machining Collect metal filings when machining metal parts. Keep separate and recycle if possible. Prevent metal filings from falling into a storm sewer drain.

Refrigerants Recover or recycle refrigerants during the service and disposal of motor vehicle air conditioners and refrigeration equipment. It is not allowable to knowingly vent refrigerants to the atmosphere. Recovering or recycling during servicing must be performed by an EPA-certified technician using certified equipment and following specified procedures.

Solvents Replace hazardous chemicals with less toxic alternatives that have equal performance. For example, substitute water-based cleaning solvents for petroleum-based sol-

vent degreasers. To reduce the amount of solvent used when cleaning parts, use a two-stage process—dirty solvent followed by fresh solvent. Hire a hazardous waste management service to clean and recycle solvents. (Some spent solvents must be disposed of as hazardous waste unless they are recycled properly.) Store solvents in closed containers to prevent evaporation. Evaporation of solvents contributes to ozone depletion and smog formation. In addition, the residue from evaporation must be treated as a hazardous waste. Properly label spent solvents and store on drip pans or in diked areas and only with compatible materials.

Containers Cap, label, cover, and properly store above ground and outdoors any liquid containers and small tanks within a diked area and on a paved impermeable surface to prevent spills from running into surface or ground water.

Other solids Store materials such as scrap metal, old machine parts, and worn tires under a roof or tarpaulin to protect them from the elements and to prevent potentially contaminated runoff. Consider recycling tires by retreading them.

Liquid recycling Collect and recycle coolants from radiators. Store transmission fluids, brake fluids, and solvents containing chlorinated hydrocarbons separately, and recycle or dispose of them properly.

Shop towels or rags Keep waste towels in a closed container marked "Contaminated Shop Towels Only." To reduce costs and liabilities associated with disposal of used towels, which can be classified as hazardous wastes, investigate using a laundry service that is able to treat the wastewater generated from cleaning the towels.

Waste storage Always keep hazardous waste separate, properly labeled, and sealed in the recommended containers. The storage area should be covered and may need to be fenced and locked if vandalism could be a problem. Select a licensed hazardous waste hauler after seeking recommendations and reviewing the firm's permits and authorizations.

NATEF TASK LIST FOR AUTOMATIC TRANSMISSIONS

A. General Transmission and Transaxle Diagnosis

A.1.   Complete work order to include customer information, vehicle identifying information, customer concern, related service history, cause, and correction.   Priority Rating 1

A.2.   Identify and interpret transmission concern; assure proper engine operation; determine necessary action.   Priority Rating 1

A.3.   Research applicable vehicle and service information, such as transmission/transaxle system operation, fluid type, vehicle service history, service precautions, and technical service bulletins.   Priority Rating 1

A.4.   Locate and interpret vehicle and major component identification numbers (VIN, vehicle certification labels, calibration decals).   Priority Rating 1

A.5.   Diagnose fluid loss and condition concerns; check fluid level on transmissions with and without dip-stick; determine necessary action.   Priority Rating 1

A.6.   Perform pressure tests; determine necessary action.   Priority Rating 1

A.7.   Perform stall test; determine necessary action.   Priority Rating 3

A.8.   Perform lock-up converter system tests; determine necessary action.   Priority Rating 3

A.9.   Diagnose mechanical and vacuum control system concerns; determine necessary action.   Priority Rating 2

A.10.   Diagnose noise and vibration concerns; determine necessary action.   Priority Rating 2

A.11.   Diagnose transmission/transaxle gear reduction/multiplication concerns using driving, driven, and held member (power flow) principles.   Priority Rating 1

A.12.   Diagnose pressure concerns in the transmission using hydraulic principles (Pascal's Law).   Priority Rating 2

A.13.   Diagnose electrical/electronic concerns using principals of electricity (Ohm's Law).   Priority Rating 2

B. Transmission and Transaxle Maintenance and Adjustment

B.1.   Inspect, adjust or replace throttle (TV) linkages or cables; manual shift linkages or cables; transmission range sensor; check gear select indicator (as applicable).   Priority Rating 1

B.2.   Service transmission; perform visual inspection; replace fluids and filters.   Priority Rating 1

C. In-Vehicle Transmission and Transaxle Repair

C.1.   Inspect, adjust or replace (as applicable) vacuum modulator; inspect and repair or replace lines and hoses.   Priority Rating 3

C.2.   Inspect, repair, and replace governor assembly.   Priority Rating 3

C.3.   Inspect and replace external seals and gaskets.   Priority Rating 2

C.4.   Inspect extension housing, bushings, and seals; perform necessary action.   Priority Rating 3

C.5.   Inspect and replace speedometer drive gear, driven gear, vehicle speed sensor (VSS), and retainers.   Priority Rating 2

C.6.   Diagnose electronic transmission control systems using a scan tool; determine necessary action.   Priority Rating 1

C.7.   Inspect, replace, and align powertrain mounts.   Priority Rating 2

D. Off-Vehicle Transmission and Transaxle Repair

1. Removal, Disassembly, and Reinstallation

D.1.1.   Remove and reinstall transmission and torque converter (rear-wheel drive).   Priority Rating 2

D.1.2.   Remove and reinstall transaxle and torque converter assembly.   Priority Rating 1

D.1.3.   Disassemble, clean, and inspect transmission/transaxle.   Priority Rating 1

D.1.4.   Inspect, measure, clean, and replace valve body (includes surfaces, bores, springs, valves, sleeves, retainers, brackets, check-balls, screens, spacers, and gaskets).   Priority Rating 2

D.1.5. Inspect servo bore, piston, seals, pin, spring, and retainers;
        determine necessary action.   Priority Rating 3
D.1.6. Inspect accumulator bore, piston, seals, spring, and retainer;
        determine necessary action.   Priority Rating 3
D.1.7. Assemble transmission/transaxle.   Priority Rating 1
D.1.8. Inspect, leak test, and flush cooler, lines, and fittings.   Priority Rating 1

2. *Oil Pump and Converter*
   D.2.1. Inspect converter flex plate, attaching parts, pilot, pump drive,
           and seal areas.   Priority Rating 2
   D.2.2. Measure torque converter endplay and check for interference;
           check stator clutch.   Priority Rating 2
   D.2.3. Inspect, measure, and replace oil pump assembly and components. Priority Rating 1

3. *Gear Train, Shafts, Bushings, and Case*
   D.3.1. Measure endplay or preload; determine necessary action.   Priority Rating 1
   D.3.2. Inspect, measure, and replace thrust washers and bearings.   Priority Rating 2
   D.3.3. Inspect oil delivery seal rings, ring grooves, and sealing
           surface areas.   Priority Rating 2
   D.3.4. Inspect bushings; determine necessary action.   Priority Rating 2
   D.3.5. Inspect and measure planetary gear assembly (includes sun gear,
           ring gear, thrust washers, planetary gears, and carrier assembly);
           determine necessary action.   Priority Rating 2
   D.3.6. Inspect case bores, passages, bushings, vents, and mating surfaces;
           determine necessary action.   Priority Rating 2
   D.3.7. Inspect transaxle drive, link chains, sprockets, gears, bearings,
           and bushings; perform necessary action.   Priority Rating 2
   D.3.8. Inspect, measure, repair, adjus,t or replace transaxle final drive
           components.   Priority Rating 2
   D.3.9. Inspect and reinstall parking pawl, shaft, spring, and retainer;
           determine necessary action.   Priority Rating 3

4. *Friction and Reaction Units*
   D.4.1. Inspect clutch drum, piston, check-balls, springs, retainers, seals,
           and friction and pressure plates; determine necessary action. Priority Rating 2
   D.4.2. Measure clutch pack clearance; determine necessary action.   Priority Rating 1
   D.4.3. Air test the operation of clutch and servo assemblies.   Priority Rating 1
   D.4.4. Inspect roller and sprag clutch, races, rollers, sprags, springs,
           cages, and retainers; determine necessary action.   Priority Rating 1
   D.4.5. Inspect bands and drums; determine necessary action.   Priority Rating 2

DEFINITION OF TERMS USED IN THE TASK LIST

To clarify the intent of these tasks, NATEF has defined some of the terms used in the task list. To get a good understanding of what the task includes, refer to this glossary while reading the task list.

adjust	To bring components to specified operational settings.
air test	To use air pressure to determine proper action of components.
align	To bring to precise alignment or relative position of components.
assemble (reassemble)	To fit together the components of a device.
check	To verify condition by performing an operational or comparative examination.
clean	To rid components of extraneous matter for the purpose of reconditioning, repairing, measuring, and reassembling.
determine	To establish the procedure to be used to effect the necessary repair.
determine necessary action	Indicates that the diagnostic routine(s) is the primary emphasis of a task. The student is required to perform the diagnostic steps and

communicate the diagnostic outcomes and corrective actions required to address the concern or problem. The training program determines the communication method (worksheet, test, verbal communication, or other means deemed appropriate) and whether the corrective procedures for these tasks are actually performed.

diagnose
To locate the root cause or nature of a problem by using the specified procedure.

disassemble
To separate a component's parts as a preparation for cleaning, inspection, or service.

drain
To use gravity to empty a container.

fill (refill)
To bring fluid level to a specified point or volume.

find
To locate a particular problem, such as shorts, grounds, or opens, in an electrical circuit.

flush
To use fluid to clean an internal system.

identify
To establish the identity of a vehicle or component before service; to determine the nature or degree of a problem.

inspect
(See *check*)

install (reinstall)
To place a component in its proper position in a system.

leak test
To locate the source of leaks in a component or system.

listen
To use audible clues in the diagnostic process; to hear the customer's description of a problem.

locate
Determine or establish a specific spot or area.

measure
To compare existing dimensions to specified dimensions by the use of calibrated instruments and gauges.

on-board diagnostics (OBD)
A diagnostic system contained in the Powertrain Control Module (PCM), which monitors computer inputs and outputs for failures. OBD II is an industry-standard, second generation OBD system that monitors emissions control systems for degradation as well as failures.

perform
To accomplish a procedure in accordance with established methods and standards.

perform necessary action
Indicates that the student is to perform the diagnostic routine(s) and perform the corrective action item. Where various scenarios (conditions or situations) are presented in a single task, at least one of the scenarios must be accomplished.

pressure test
To use air or fluid pressure to determine the integrity, condition, or operation of a component or system.

priority ratings
Indicates the minimum percentage of tasks, by area, a program must include in its curriculum in order to be certified in that area.

reassemble
(See *assemble*)

remove
To disconnect and separate a component from a system.

repair
To restore a malfunctioning component or system to operating condition.

replace
To exchange an unserviceable component with a new or rebuilt component; to reinstall a component.

reset
(See *set*)

select	To choose the correct part or setting during assembly or adjustment.
service	To perform a specified procedure when called for in the owner's manual or service manual.
set	To adjust a variable component to a given, usually initial, specification.
test	To verify condition through the use of meters, gauges, or instruments.
torque	To tighten a fastener to specified degree or tightness (in a given order or pattern if multiple fasteners are involved on a single component).

TOOLS AND EQUIPMENT

Many different tools and testing and measuring equipment are used to service automatic transmissions. NATEF has identified many of these and has said that an automatic transmission technician must know what they are and how and when to use them. The tools and equipment listed by NATEF are covered in the following discussion. Also included are the tools and equipment you will use while completing the job sheets. Although you will be using common hand tools, they are not part of this discussion. You should already know what they are and how to use and care for them.

Scan Tools

The introduction of computer-controlled systems brought with it the need for tools capable of troubleshooting electronic control systems. A variety of computer scan tools are available today that do just that. A scan tool is a microprocessor designed to communicate with the vehicle's computer. Connected to the computer through diagnostic connectors, a scan tool can access trouble codes, run tests to check system operations, and monitor the activity of the system. Trouble codes and test results are displayed on an LED screen or printed out on the scanner printer.

Scan tools retrieve fault codes from a computer's memory and digitally display these codes on the tool. A scan tool may also perform many other diagnostic functions, depending on the year and make of the vehicle. Most aftermarket scan tools have removable modules that are updated each year. These modules are designed to test the computer systems on various makes of vehicles. For example, some scan testers have a 3-in-1 module that tests the computer systems on Chrysler, Ford, and General Motors vehicles. A 10-in-1 module is also available to diagnose computer systems on vehicles imported by 10 different manufacturers. These modules plug into the scan tool.

Scan tools are capable of testing many onboard computer systems, such as transmission controls, engine computers, antilock brake computers, air bag computers, and suspension computers, depending on the year and make of the vehicle and the type of scan tester. In many cases, the technician must select the computer system to be tested with the scanner after it has been connected to the vehicle.

The scan tool is connected to specific diagnostic connectors on various vehicles. Most manufacturers have one diagnostic connector. It connects the data wire from each onboard computer to a specific terminal in this connector. Other vehicle manufacturers have several different diagnostic connectors on each vehicle, each of which may be linked to one or more onboard computers. A set of connectors is supplied with the scanner to allow tester connection to various diagnostic connectors on different vehicles.

The scanner must be programmed for the model year, make of vehicle, and type of engine. With some scan tools, this selection is made by pressing the appropriate buttons on the tester, as directed by the digital tester display. On other scan testers, the appropriate memory card must be installed in the tester for the vehicle being tested. Some scan testers have a built-in printer to print test results, whereas other scan testers may be connected to an external printer.

As automotive computer systems become more complex, the diagnostic capabilities of scan testers continue to expand. Many scan testers now have the capability to store, or freeze, data into the tester during a road test and then play it back when the vehicle is returned to the shop.

Some scan testers now display diagnostic information based on the fault code in the computer memory. Service bulletins published by the manufacturer of the scan tester may be indexed by the tester after the vehicle information is entered in the tester. Other scan testers will display sensor specifications for the vehicle being tested.

Trouble codes are set off by the vehicle's computer only when a voltage signal is entirely out of its normal range. The codes help technicians identify the cause of the problem. If a signal is within its normal range but still not correct, the vehicle's computer does not display a trouble code, but a problem still exists. To help identify this type of problem, most manufacturers recommend that the signals to and from the computer be carefully looked at. This is done through the use of a scan tool or breakout box. A breakout box allows the technician to check voltage and resistance readings between specific points within the computer's wiring harness.

With On Board Diagnostic II (OBD-II), the diagnostic connectors are located in the same place on all vehicles. Any scan tools designed for OBD-II work on all OBD-II systems; therefore there is no need to have designated scan tools or cartridges. The OBD-II scan tool can run diagnostic tests on all systems and has "freeze frame" capabilities.

Vacuum Gauge

Measuring intake manifold vacuum is another way to diagnose the condition of an engine. This is important when diagnosing automatic transmissions. A transmission responds to engine load, and engine load affects engine vacuum. As load increases, engine vacuum decreases. This means that a poorly running engine will be seen by the transmission as always having a load on it, which results in poor shift quality and inefficient shift timing.

Manifold vacuum is tested with a vacuum gauge. Vacuum is formed on a piston's intake stroke. As the piston moves down, it lowers the pressure of the air in the cylinder—if the cylinder is sealed. This lower cylinder pressure is called engine vacuum. If there is a leak, atmospheric pressure forces air into the cylinder and the resultant pressure is not as low. The reason atmospheric pressure enters is simply that whenever low and high pressures exist, high pressure always moves toward the low pressure.

Vacuum is measured in inches of mercury (in./Hg) and in kiloPascals (kPa) or millimeters of mercury (mm/Hg).

To measure vacuum, a flexible hose on the vacuum gauge is connected to a source of manifold vacuum, either on the manifold or at a point below the throttle plates. Sometimes this requires removing a plug from the manifold and installing a special fitting.

The test is made with the engine cranking or running. A good vacuum reading is typically at least 16 in./Hg. However, a reading of 15 to 20 in./Hg (50 to 65 kPa) is normally acceptable. Since the intake stroke of each cylinder occurs at a different time, the production of vacuum occurs in pulses. If the amount of vacuum produced by each cylinder is the same, the vacuum gauge will show a steady reading. If one or more cylinders are producing different amounts of vacuum, the gauge will show a fluctuating reading.

Hydraulic Pressure Gauge Set

A common diagnostic tool for automatic transmissions is a hydraulic pressure gauge. A pressure gauge measures pressure in pounds per square inch (psi) and/or kPa. The gauge is normally part of a kit that contains various fittings and adapters.

Portable Crane

To remove and install a transmission, the engine is often moved out of the vehicle with the transmission. To remove or install an engine, a portable crane is used. A crane uses hydraulic pressure that is converted to a mechanical advantage and lifts the engine from the vehicle. To lift an engine, attach a pulling sling or chain to the engine. Some engines have eye plates for use in lifting. If they are not available, the sling must be bolted to the engine. The sling attaching bolts must be large enough to support the engine and must thread into the block at least 1.5 times the bolt diameter. Connect the crane to the chain. Raise the engine slightly and make sure the sling attachments are secure. Carefully lift the engine out of its compartment.

Lower the engine close to the floor so that the transmission and torque converter can be removed from the engine if necessary.

Transmission Jacks

Transmission jacks (Figure 2) are designed to help you while removing a transmission from under the vehicle. The weight of the transmission makes it difficult and unsafe to remove it without assistance or the use of a transmission jack. These jacks fit under the transmission and are typically

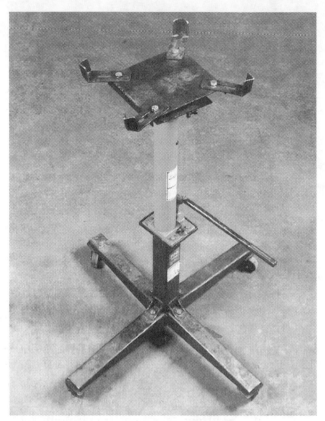

Figure 2   A transmission jack.

equipped with hold-down chains. These chains are used to secure the transmission to the jack, and the transmission's weight rests on the jack's saddle.

Transmission jacks are available in two basic styles. One is used when the vehicle is raised by a hydraulic jack and is set on jack stands. The other style is used when the vehicle is raised on a lift.

Transaxle Removal and Installation Equipment

Removal and replacement of transversely mounted engines may require other tools. The engines of some FWD vehicles are removed by lifting them from the top. Others must be removed from the bottom, which requires different equipment. Make sure you follow the instructions given by the manufacturer and use the appropriate tools and equipment. The required equipment varies with manufacturer and vehicle model but most accomplish the same thing.

To remove the engine and transmission from under the vehicle, the vehicle must be raised. A crane or support fixture (or both) is used to hold the engine and transaxle assembly in place while the assembly is being readied for removal. When everything is set for removal of the assembly, the

crane is used to lower the assembly onto a cradle. The cradle is similar to a hydraulic floor jack and is used to lower the assembly further so it can be rolled out from under the vehicle. The transaxle can be separated from the engine once it has been removed from the vehicle.

When the transaxle is removed as a single unit, the engine must be supported while it is in the vehicle before, during, and after transaxle removal. Special fixtures mount to the vehicle's upper frame or suspension parts (Figure 3). These supports have a bracket that is attached to the engine. With the bracket in place, the engine's weight is now on the support fixture, and the transmission can be removed.

Transmission/Transaxle Holding Fixtures

Special holding fixtures should be used to support the transmission or transaxle after it has been removed from the vehicle. These holding fixtures may be standalone units or bench mounted; they allow the transmission to be easily repositioned during repair work.

Machinist's Rule

A machinist's rule is very much like an ordinary ruler. Each edge of this measuring tool is divided into increments based on a different scale. A typical machinist's rule based on the USCS measurement may have scales based on 1/8-, 1/16-, 1/32-, and 1/64-inch intervals. Of course, metric machinist rules are also available. Metric rules are usually divided into 0.5-mm and 1-mm increments.

Some machinist's rules are based on decimal intervals. These are typically divided into 1/10-,

Figure 3   A transverse engine support bar.

1/100-, and 1/1,000-inch (0.1, 0.01, and 0.001) increments. Decimal machinist's rules are very helpful when measuring dimensions that are specified in decimals, so that you do not have to convert fractions to decimals.

Micrometers

A micrometer is used to measure linear outside and inside dimensions. Both outside and inside micrometers are calibrated and read in the same manner. The major components and markings of a micrometer include the frame, anvil, spindle, lock nut, sleeve, sleeve numbers, sleeve long line, thimble marks, thimble, and ratchet. Micrometers are calibrated in either inch or metric graduations and are available in a range of sizes.

To use and read a micrometer, choose the appropriate size for the object being measured (Figure 4). Typically they measure an inch; therefore, the range covered by one size of micrometer would be from 0 to 1 inch, another would measure 1 to 2 inches, and so on.

Open the jaws of the micrometer and slip the object between the spindle and anvil. While holding the object against the anvil, turn the thimble using your thumb and forefinger until the spindle contacts the object. Never clamp the micrometer tightly. Use only enough pressure on the thimble to allow the work to just fit between the anvil and spindle. To get accurate readings, you should slip the micrometer back and forth over the object until you feel a very light resistance while rocking the tool from side to side to make certain the spindle cannot be closed any further. When a satisfactory adjustment has been made, lock the micrometer. Read the measurement scale.

The graduations on the sleeve each represent 0.025 inch. To read a measurement on a microm-

eter, begin by counting the visible lines on the sleeve and multiplying them by 0.025. The graduations on the thimble assembly define the area between the lines on the sleeve. The number indicated on the thimble is added to the measurement shown on the sleeve. This sum is the dimension of the object.

Micrometers are available to measure in 0.0001 (ten-thousandths) of an inch. Use this type of micrometer if the specifications call for this much accuracy.

A metric micrometer is read in the same way, except that the graduations are expressed in the metric system of measurement. Each number on the sleeve represents 5 millimeters (mm) or 0.005 meter (m). Each of the 10 equal spaces between each number, with index lines alternating above and below the horizontal line, represents 0.5 mm, or five-tenths of a millimeter. Therefore, one revolution of the thimble changes the reading one space on the sleeve scale, or 0.5 mm. The beveled edge of the thimble is divided into 50 equal divisions with every fifth line numbered: 0, 5, 10, . . . , 45. Since one complete revolution of the thimble advances the spindle 0.5 mm, each graduation on the thimble is equal to one-hundredth of a millimeter. As with the inch-graduated micrometer, the separate readings are added together to obtain the total reading.

Some technicians use a digital micrometer, which is easier to read. These tools do not have the various scales; rather, the measurement is displayed and read directly off the micrometer.

Inside micrometers can be used to measure the inside diameter of a bore. To do this, place the tool inside the bore and extend the measuring surfaces until each end touches the bore's surface. If the bore is large, it might be necessary to use an extension rod to increase the micrometer's range. These extension rods come in various lengths. The inside micrometer is read in the same manner as an outside micrometer.

A depth micrometer (Figure 5) is used to measure the distance between two parallel surfaces. The sleeves, thimbles, and ratchet screws operate in the same way as other micrometers. Depth micrometers are read in the same way as other micrometers.

If a depth micrometer is used with a gauge bar, it is important to keep both the bar and the micrometer from rocking. Any movement of either part results in an inaccurate measurement.

Figure 4   An outside micrometer.

Figure 5 A depth micrometer.

Telescoping Gauge

Telescoping gauges are used for measuring bore diameters and other clearances. They may also be called snap gauges. They are available in sizes ranging from fractions of an inch through 6 inches. Each gauge consists of two telescoping plungers, a handle, and a lock screw. Snap gauges are normally used with an outside micrometer.

To use the telescoping gauge, insert it into the bore and loosen the lock screw. This allows the plungers to snap against the bore. Once the plungers have expanded, tighten the lock screw. Then remove the gauge and measure the expanse with a micrometer.

Small Hole Gauge

A small hole or ball gauge works just like a telescoping gauge except that it is designed for small bores. After it is placed into the bore and expanded, it is removed and measured with a micrometer. Like the telescoping gauge, the small hole gauge consists of a lock, a handle, and an expanding end. The end expands or retracts by turning the gauge handle.

Feeler Gauge

A feeler gauge is a thin strip of metal or plastic of known and closely controlled thickness. Several of these strips are often assembled together as a feeler gauge set that looks like a pocketknife. The desired thickness gauge can be pivoted away from others for convenient use. A feeler gauge set usually contains strips or leaves of 0.002- to 0.010-inch thickness (in steps of 0.001-inch) and leaves of 0.012- to 0.024-inch thickness (in steps of 0.002-inch).

A feeler gauge can be used by itself to measure piston clearance, fluid pump clearances, and other distances. It can also be used with a precision straightedge to check the flatness of a sealing surface.

Straightedge

A straightedge is no more than a flat bar machined to be totally flat and straight; to be effective, it must be flat and straight. Any surface that should be flat can be checked with a straightedge and feeler gauge set. The straightedge is placed across and at angles to the surface. At any low points on the surface, a feeler gauge can be placed between the straightedge and the surface. The size gauge that fills in the gap is the amount of warpage or distortion.

Dial Indicator

The dial indicator is calibrated in 0.001-inch (one-thousandth-inch) increments. Metric dial indicators are also available. Both types are used to measure movement. Common uses of the dial indicator include measuring endplay (Figure 6), clutch pack clearances, and flexplate runout.

To use a dial indicator, position the indicator rod against the object to be measured. Then push the indicator toward the work until the indicator needle travels far enough around the gauge face to permit movement to be read in either direction. Zero the indicator needle on the gauge. Move the object in the direction required while observing the needle of the gauge. Always be sure the range of the dial indicator is sufficient to allow the amount of movement required by the measuring procedure. For example,

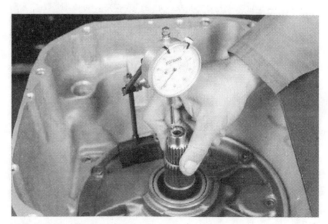

Figure 6 A dial indicator with a holding fixture.

never use a 1-inch indicator on a component that will move 2 inches.

Torque-Indicating Wrench

Torque is the twisting force used to turn a fastener against the friction between the threads and between the head of the fastener and the surface of the component. The fact that practically every vehicle and engine manufacturer publishes a list of torque recommendations is ample proof of the importance of using proper amounts of torque when tightening nuts or bolts. The amount of torque applied to a fastener is measured with a torque-indicating or torque wrench.

There are three basic types of torque indicating wrenches available. The following three types are available with pounds-per-inch (Figure 7) and pounds-per-foot increments: a beam torque wrench, which has a beam pointing to the torque reading; a click-type torque wrench, in which the desired torque reading is set on the handle (when the torque reaches that level, the wrench clicks); and a dial torque wrench, which has a dial that indicates the torque exerted on the wrench. Some designs of this type of torque wrench have a light or buzzer that turns on when the desired torque is reached.

Blowgun

Blowguns are used for air testing components (Figure 8) and blowing off parts during cleaning. Never point a blowgun at yourself or someone else. A blowgun snaps into one end of an air hose and directs airflow when a button is pressed. Always use an OSHA-approved air blowgun. Before using a blowgun, be sure it has not been modified to eliminate air-bleed holes on the side.

Figure 8   An air nozzle used while air testing.

Gear and Bearing Pullers

Many tools are designed for a specific purpose. An example of a special tool is a gear and bearing puller. Many gears and bearings have a slight interference fit (press fit) when they are installed on a shaft or in a housing. Something that has a press fit has an interference fit; for example, the inside diameter of a bore is 0.001 inch smaller than the outside diameter of a shaft, so when the shaft is fitted into the bore, it must be pressed in to overcome the 0.001-inch interference fit. This press fit prevents the parts from moving on each other. These gears and bearings must be removed carefully to prevent damage to the gears, bearings, or shafts. Prying or hammering can break or bind the parts. A puller with the proper jaws and adapters should be used to remove gears and bearings. With the proper puller, the force required to remove a gear or bearing can be applied with a slight and steady motion.

Bushing and Seal Pullers and Drivers

Another commonly used group of special tools are the various designs of bushing and seal drivers (Figure 9) and pullers. Pullers are either a threaded or slide hammer type tool. Always make sure you use the correct tool for the job because bushings and seals are easily damaged if the wrong tool or procedure is used. Car manufacturers and specialty tool companies work closely together to design and manufacture special tools required to repair cars. Most of these special tools are listed in the appropriate service manuals.

Retaining Ring Pliers

An automatic transmission technician runs into many different styles and sizes of retaining rings

Figure 7   A pound per inch torque wrench.

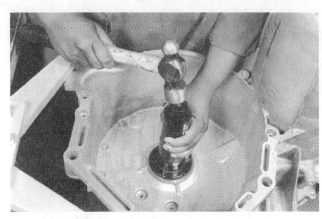

Figure 9 An example of a seal driver.

used to hold subassemblies together or keep them in a fixed location (Figure 10). Using the correct tool to remove and install these rings is the only safe way to work with them. All automatic transmission technicians should have an assortment of retaining ring pliers.

Special Tool Sets

Vehicle manufacturers and specialty tool companies work closely together to design and manufacture special tools required to repair transmissions. Most of these special tools are listed in the appropriate service manuals and are part of each manufacturer's Essential Tool Kit.

Service Manuals

Perhaps the most important tools you will use are service manuals. There is no way a technician can remember all the procedures and specifications needed to repair all vehicles. Thus, a good technician relies on service manuals and other information sources. Good information plus knowledge

Figure 10 Snap ring pliers.

allows a technician to fix a problem with the least frustration and at the lowest expense to the customer.

To obtain the correct transmission specifications and other information, you must first identify the transmission you are working on. The best source for positive identification is the Vehicle Identification Number (VIN). The transmission code can be interpreted through information given in the service manual, which may also help you identify the transmission through appearance, casting numbers, or markings on the housing.

The primary source of repair and specification information for any car, van, or truck is the manufacturer. The manufacturer publishes service manuals each year for every vehicle built. Because of the enormous amount of information, some manufacturers publish more than one manual per year per car model. They are typically divided into sections based on the major systems of the vehicle. In the case of transmissions, there is a section for each transmission that may be in the vehicle. Manufacturers' manuals cover all repairs, adjustments, specifications, detailed diagnostic procedures, and special tools required.

Since many technical changes occur on specific vehicles each year, manufacturers' service manuals need to be updated constantly. Updates are published as service bulletins (often referred to as technical service bulletins, or TSBs) that show the changes in specifications and repair procedures during the model year. These changes do not appear in the service manual until the next year. The car manufacturer provides these bulletins to dealers and repair facilities on a regular basis.

Service manuals are also published by independent companies rather than the manufacturers. However, they pay for and get most of their information from the carmakers. They contain component information, diagnostic steps, repair procedures, and specifications for several car makes in one book. Information is usually condensed and is more general in nature than the manufacturer's manuals. The condensed format allows for more coverage in less space and, therefore, is not always specific. They may also contain several years of models as well as several car makes in one book.

Many of the larger parts manufacturers have excellent guides on the various parts they manufacture or supply. They also provide updated service bulletins on their products. Other sources for up-to-date technical information are trade magazines and trade associations.

The same information that is available in service manuals is now commonly found electronically on compact disks (CD-ROMs), digital video disks (DVDs), and the internet. A single compact disk can hold a quarter million pages of text, eliminating the need for a huge library to contain all of the printed manuals. Using electronics to find information is also easier and quicker. The disks are normally updated quarterly and not only contain the most recent service bulletins but also engineering and field service fixes. DVDs can hold more information than CDs; therefore, fewer disks are needed with systems that use DVDs. The CDs and DVDs are inserted into a computer. All a technician needs to do is enter vehicle information and then move to the appropriate part or system. The appropriate information will then appear on the computer's screen. Online data can be updated instantly and requires no space for physical storage. These systems are easy to use and the information is quickly accessed and displayed. The computer's keyword, mouse, and/or light pen are used to make selections from the screen's menu. Once the information is retrieved, a technician can read it off the screen or print it out and take it to the service bay.

CROSS REFERENCE GUIDE

NATEF Task	Job Sheet	NATEF Task	Job Sheet
A.1	1	D.1.3	19
A.2	2	D.1.4	20
A.3	3	D.1.5	21
A.4	3	D.1.6	21
A.5	4	D.1.7	22
A.6	5	D.1.8.	23
A.7	6	D.2.1	24
A.8	7	D.2.2	25
A.9	8	D.2.3	26
A.10	2	D.3.1	22
A.11	9	D.3.2	27
A.12	10	D.3.3	28
A.13	8	D.3.4	27
B.1	11	D.3.5	29
B.2	12	D.3.6	30
C.1	8	D.3.7	31
C.2	13	D.3.8	32
C.3	14	D.3.9	33
C.4	15	D.4.1	34
C.5	15	D.4.2	35
C.6	16	D.4.3	35
C.7	17	D.4.4	34
D.1.1	18	D.4.5	36
D.1.2	18		

JOB SHEETS

AUTOMATIC TRANSMISSIONS AND TRANSAXLES
JOB SHEET 1

Filling out a Work Order

Name _____ Station _____ Date _____

NATEF Correlation:

This Job Sheet addresses the following NATEF task:

A.1. Complete work order to include customer information, vehicle identifying information, customer concern, related service history, cause, and correction.

Objective

Upon completion of this job sheet, you will be able to prepare a service work order based on customer input, vehicle information, and service history.

Tools and Materials

An assigned vehicle or the vehicle of your choice

Service work order or computer-based shop management package

Parts and labor guide

Work Order Source: Describe the system used to complete the work order. If a paper repair order is being used, describe the source. _____

_____

PROCEDURE

1. Prepare the shop management software for entering a new work order or obtain a blank paper work order.   Task Completed ☐

2. Enter customer information, including name, address, and phone numbers onto the work order.   Task Completed ☐

3. Locate and record the vehicle's VIN.   Task Completed ☐

4. Enter the necessary vehicle information, including year, make, model, engine type and size, transmission type, license number, and odometer reading.   Task Completed ☐

5. Does the VIN verify that the information about the vehicle is correct?

   _____

6. Normally, you would interview the customer to identify his or her concerns. However to complete this job sheet, assume the customer desires to have the fluid and filter changed in the transmission.   Task Completed ☐

7. The history of service to the vehicle can often help diagnose problems as well as indicate possible premature part failure. Gathering this information from the customer can provide some of this information. For this job sheet assume the vehicle has not had a similar problem and was not recently involved in a collision. Service history is further obtained by searching files based on customer name, VIN, and license number. Check the files for any related service work.   Task Completed ☐

8. Search for technical service bulletins on this vehicle that may relate to the customer's concern.   Task Completed ☐

9. Based on the customer's concern, service history, TSBs, and your knowledge, what is the likely cause of this concern?

   _____

   _____

10. Enter this information onto the work order.   Task Completed ☐

11. Prepare to make a repair cost estimate for the customer. Identify all parts that may need to be replaced to correct the concern. List these here.

    _____

    _____

12. Describe the task(s) that will be necessary to replace the part.

    _____

    _____

13. Using the parts and labor guide, locate the cost of the parts that will be replaced and enter the cost of each item onto the work order at the appropriate place for creating an estimate.   Task Completed ☐

14. Now, locate the flat rate time for work required to correct the concern. List each task and with its flat rate time.

    _____

    _____

15. Multiply the time for each task by the shop's hourly rate and enter the cost of each item onto the work order at the appropriate place for creating an estimate.

Task Completed ☐

16. Many shops have a standard amount they charge each customer for shop supplies and waste disposal. For this job sheet, use an amount of ten dollars for shop supplies.

Task Completed ☐

17. Add the total costs and insert the sum as the subtotal of the estimate.

Task Completed ☐

18. Taxes must be included in the estimate. What is the sales tax rate and does it apply to both parts and labor, or just one of these?

_____

_____

19. Enter the appropriate amount of taxes to the estimate, than add this to the subtotal. The end result is the estimate to give the customer.

Task Completed ☐

20. By law, how accurate must your estimate be?

_____

_____

21. Generally speaking, the work order is complete and is ready for the customer's signature. However, some businesses require additional information; make sure you enter that information to the work order. On the work order there is a legal statement that defines what the customer is agreeing to. Briefly describe the contents of that statement.

_____

_____

Problems Encountered

_____

_____

_____

Instructor's Comments

_____

_____

_____

_____

AUTOMATIC TRANSMISSIONS AND TRANSAXLES JOB SHEET 2

Road Test a Vehicle to Check the Operation of an Automatic Transmission

Name _____ Station _____ Date _____

NATEF Correlation

This Job Sheet addresses the following NATEF tasks:

A.2.   Identify and interpret transmission concern; ensure proper engine operation; determine necessary action.

A.10. Diagnose noise and vibration concerns; determine necessary action.

Objective

Upon completion of this job sheet, you will be able to identify and interpret transmission concerns, ensure proper engine operation, and diagnose noise and vibration problems.

Tools and Materials

Service manual

Clean rag

Paper and pencil

Describe the vehicle being worked on:

Year _____ Make _____ Model _____

VIN _____ Engine type and size _____

Model and type of transmission: _____

PROCEDURE

1. Park the vehicle on a level surface.   Task Completed ☐

2. Wipe all dirt off the protective disc and the dipstick handle.   Task Completed ☐

3. Start the engine and allow it to reach operating temperature.   Task Completed ☐

4. Remove the dipstick and wipe it clean with a lint-free cloth or paper towel.   Task Completed ☐

5. Reinsert the dipstick, remove it again, and record the fluid's level:

   _____

6. Describe the condition of the fluid and tell what the fluid indicates.

   _____

   _____

7. Find and duplicate the chart from the service manual that shows the band and clutch application for the different gear selector positions. Using these charts will greatly simplify your diagnosis of a problem. It is also wise to have paper and a pencil to jot down notes about the operation of the transmission.

Task Completed ☐

8. Inspect the transmission for leaks and other signs of wear or damage. Comments:

_____

_____

9. Drive the vehicle at normal speeds to warm up the engine and transmission. Describe the overall behavior of the transmission and torque converter.

_____

_____

10. Place the shift selector into the Drive or Overdrive position. Allow the transmission to shift through all of its gears. Describe the operation of the transmission and torque converter. If something sounded or felt abnormal during any of the shifts or while operating in a particular gear, refer to the band and clutch application chart to identify possible causes of the problem.

_____

_____

11. Force the transmission to downshift and record the quality of that downshift and the speed at which it downshifted.

_____

_____

12. Manually cause the transmission to downshift. Record the quality of that shift.

_____

_____

13. Stop the vehicle and put the shift selector in the lowest gear. Begin driving and manually shift the transmission through the forward gears. Record the quality of those shifts and your conclusions drawn from this drive. Again, if a problem was evident, refer to the band and clutch application chart to identify possible causes.

_____

_____

14. Stop the vehicle and place the gear selector into reverse. Describe the feel of that shift.

   _____

   _____

15. With the vehicle stopped, place the gear selector into park. Describe the feel of that shift.

   _____

   _____

16. What are your recommendations for this transmission?

   _____

   _____

   _____

Problems Encountered

_____

_____

_____

Instructor's Comments

_____

_____

_____

AUTOMATIC TRANSMISSIONS AND TRANSAXLES JOB SHEET 3

Gathering Vehicle Information

Name _____ Station _____ Date _____

NATEF Correlation

This Job Sheet addresses the following NATEF tasks:

A.3. Research applicable vehicle and service information, such as transmission/transaxle operation, fluid type, vehicle service history, service precautions, and technical service bulletins.

A.4. Locate and interpret vehicle and major component identification numbers (VIN, vehicle certification labels, calibration decals).

Objective

Upon completion of this job sheet, you will be able to gather service information about a vehicle and its automatic transmission or transaxle.

Tools and Materials

Appropriate service manuals

Computer

Protective Clothing

Goggles or safety glasses with side shields

Describe the vehicle being worked on:

Year _____ Make _____ Model _____

VIN _____

PROCEDURE

1. Using the service manual or other information source, describe what each letter and number in the VIN for this vehicle represents.

   _____

   _____

2. Locate the Vehicle Emissions Control Information (VECI) label and describe where you found it.

   _____

   _____

3. Summarize what information you found on the VECI label.

   _____

   _____

4. While looking in the engine compartment or under the vehicle, locate the identification tag on the transmission or transaxle. Describe where you found it.

   _____

   _____

5. Summarize the information contained on this label.

   _____

   _____

   _____

6. Using a service manual or electronic database, locate the information about the vehicle's automatic transmission. List the major components of the control system and describe the basic operation parameters.

   _____

   _____

   _____

   _____

   _____

7. Using a service manual or electronic database, locate and record all service precautions noted by the manufacturer regarding the automatic transmission.

   _____

   _____

   _____

   _____

8. Using the information that is available, locate and record the vehicle's service history.

   _____

   _____

   _____

   _____

9. Using the information sources that are available, summarize all Technical Service Bulletins for this vehicle that relate to the automatic transmission and its control system.

_____

_____

_____

_____

10. Refer to the service manual and identify the exact type of fluid that should be used in this transmission. What type?

_____

Problems Encountered

_____

_____

_____

Instructor's Comments

_____

_____

_____

AUTOMATIC TRANSMISSIONS AND TRANSAXLES JOB SHEET 4

Visual Inspection of an Automatic Transmission

Name _____ Station _____ Date _____

NATEF Correlation

This Job Sheet addresses the following NATEF tasks:

A.5. Diagnose fluid loss and condition concerns; check fluid level on transmission with and without dipstick; determine necessary action.

Objective

Upon completion of this job sheet, you will be able to diagnose unusual fluid usage, level, and condition concerns; inspect and replace external seals and gaskets; perform lockup converter system tests; and check torque converter and transmission cooling system for contamination.

Tools and Materials

Service manual

Flashlight

Protective Clothing

Goggles or safety glasses with side shields

Describe the vehicle being worked on:

Year _____ Make _____ Model _____

VIN _____ Engine type and size _____

Model and type of transmission: _____

PROCEDURE

1. Check the transmission housing for damage, cracks, and signs of leaks. Record your findings and recommendations:

   _____

   _____

2. Check the slip joint area in the transmission extension housing for leaks. If the transmission is a transaxle, check the area where the inner CV joint is attached to the housing. Record your findings and recommendations:

   _____

   _____

3. Check for leaks wherever something is attached to the housing. Record your findings and recommendations:

_____

_____

4. Check the shift and throttle linkages for looseness, wear, and/or damage. Record your findings and recommendations:

_____

_____

5. Check any cables for binding, wear, and/or damage. Record your findings and recommendations:

_____

_____

6. Check the condition of all cooler lines and hoses. Record your findings and recommendations:

_____

_____

7. Check the condition of the fluid; pay attention to the level, color, smell, and feel of the fluid. Record your findings and recommendations:

_____

_____

Problems Encountered

_____

_____

_____

Instructor's Comments

_____

_____

_____

AUTOMATIC TRANSMISSIONS AND TRANSAXLES JOB SHEET 5

Pressure Testing an Automatic Transmission

Name _____ Station _____ Date _____

NATEF Correlation

This Job Sheet addresses the following NATEF task:

A.6. Perform pressure tests; determine necessary action.

Objective

Upon completion of this job sheet, you will have demonstrated the ability to conduct a pressure test on a transmission.

Tools and Materials

Hoist
Tachometer
Pressure gauges with
    necessary adapters

Vacuum gauge
T-fitting and misc. hoses
Lint-free shop towels
Service manual

Protective Clothing

Goggles or safety glasses with side shields

Describe the vehicle being worked on:

Year _____ Make _____ Model _____

VIN _____ Engine type and size _____

Model and type of transmission: _____

PROCEDURE

1. List the pressure specifications with the required operating conditions:

   _____

   _____

2. Start the engine and allow it to reach normal operating temperature. Then turn off the engine.   Task Completed ☐

3. Connect the tachometer if the vehicle is not equipped with one.   Task Completed ☐

4. Have a fellow student sit in the vehicle to operate the throttle, brakes, and transmission during the test.   Task Completed ☐

5. Raise the vehicle on the hoist to a comfortable working height.   Task Completed ☐

6. Describe the location of the pressure taps on the transmission.

   _____

   _____

7. Connect the pressure gauges to the appropriate service ports.

Task Completed ☐

8. If the transmission has a vacuum modulator, use the T-fitting and vacuum hoses to connect the vacuum gauge into the modulator circuit.

Task Completed ☐

9. Start the engine. Have your helper press firmly on the brake pedal and apply the parking brake. Then the helper should move the gear selector into the first test position.

Task Completed ☐

10. Run the engine at the specified speed. Move the gear selector as required by the service manual.

Task Completed ☐

11. Observe and record the pressure and vacuum readings at the various test conditions.

_____

_____

_____

12. Turn off the engine and move the pressure gauges to the appropriate test ports for the next transmission range to be tested. Describe this location.

_____

_____

_____

13. Restart the engine and have your helper move the gear selector into the range to be tested and increase the engine's speed to the required test speed.

Task Completed ☐

14. Observe and record the pressure and vacuum readings in the various test conditions.

_____

_____

_____

15. Allow the engine to return to idle and turn it off.

Task Completed ☐

16. Repeat this sequence until all transmission ranges have been tested.

Task Completed ☐

17. Summarize the results of this test and compare them to specifications.

_____

_____

_____

_____

_____

_____

Problems Encountered

_____

_____

_____

Instructor's Comments

_____

_____

_____

AUTOMATIC TRANSMISSIONS AND TRANSAXLES JOB SHEET 6

Conduct a Stall Test

Name _____ Station _____ Date _____

NATEF Correlation

This Job Sheet addresses the following NATEF task:

A.7. Perform stall test; determine necessary action.

Objective

Upon completion of this job sheet, you will be able to conduct a stall test.

Tools and Materials

A vehicle with an automatic transmission

Tachometer

Hoist

Stethoscope

Droplight or good flashlight

Service manual

Protective Clothing

Goggles or safety glasses with side shields

Describe the vehicle being worked on:

Year _____ Make _____ Model _____

VIN _____ Engine type and size _____

PROCEDURE

1. Describe the transmission in the vehicle being tested:

   _____

   _____

2. Park the vehicle on a level surface.   Task Completed ☐

3. Describe the condition of the fluid (color, condition, and smell).

   _____

   _____

4. What is indicated by the fluid's condition?

   _____

   _____

5. Connect a tachometer to the engine. If the vehicle has one in the instrument panel, there is no need to connect another one. Explain how you connected the tachometer:

_____

_____

6. Place the tachometer on the inside of the vehicle so that it can be easily read.   Task Completed ☐

7. Mark the face of the tachometer with a grease pencil at the recommended maximum rpm for this test. What is the maximum rpm? _____

8. Check the engine's coolant level. Describe your findings.

_____

_____

9. Block the front wheels and set the parking brake.   Task Completed ☐

10. Place the transmission in park and allow the engine and transmission to warm.   Task Completed ☐

11. Place the gear selector into the gear that is recommended for this test. What is the recommended gear?_____

12. Press the throttle pedal to the floor with your right foot and firmly press the brake pedal with your left. Hold the brake pedal down.   Task Completed ☐

13. When the tachometer needle stops rising, note the engine speed and let off the throttle. What was the highest rpm during the test? _____

14. Place the gearshift into neutral and allow the engine to run at 1000 rpm for at least one minute. What is the purpose of doing this?

_____

_____

15. If noise is heard from the transmission during the stall test, raise the vehicle on a hoist.   Task Completed ☐

16. Use the stethoscope to determine if the noise is from the transmission or the torque converter.   Task Completed ☐

17. What are your conclusions from this test?

_____

_____

_____

_____

Problems Encountered

_____

_____

_____

Instructor's Comments

_____

_____

_____

AUTOMATIC TRANSMISSIONS AND TRANSAXLES JOB SHEET 7

Testing a Lock-up Converter

Name _____ Station _____ Date _____

NATEF Correlation

This Job Sheet addresses the following NATEF task:

A.8. Perform lock-up converter system tests; determine necessary action.

Objective

Upon completion of this job sheet, you will be able to test the operation of a lock-up converter.

Tools and Materials

Service manual

Scan tool

Digital multimeter

Protective Clothing

Goggles or safety glasses with side shields

Describe the vehicle being worked on:

Year _____ Make _____ Model _____

VIN _____ Engine type and size _____

PROCEDURE

1. Check with the service or owner's manual to identify the lamp that warns the driver of a transmission concern. What lamp is it?

   _____

   _____

2. Start the engine and observe the above lamp and the engine MIL. What did you observe?

   _____

   _____

3. Connect the scan tool and retrieve all DTCs. What were they?

   _____

   _____

4. If an engine related problem is suspected, it should be corrected before moving on to the torque converter tests.   Task Completed ☐

5. Operate the vehicle with the scan tool connected. Drive it at 40 mph.   Task Completed ☐

6. With the scan tool, engage the lock up converter's solenoid. Watch the tachometer while making the change. What did you observe? Engine speed should drop when the converter locks.

_____

_____

7. Disengage the clutch and observe the engine's speed. What did you observe?

_____

_____

8. The action of the lock-up converter can also be checked with the vehicle stopped and the engine running. Place the shifter into the "drive" position.

Task Completed ☐

9. With the scan tool, engage the lock-up solenoid. Describe what happened.

_____

_____

10. If there was no change in engine speed, what is evident?

_____

_____

11. The lock-up solenoid can also be checked with the ignition on but the engine off. Engage the solenoid with the scan tool and listen. Did you hear a click?

_____

_____

12. If a problem with the solenoid is suspected, remove it from the transmission for testing.

Task Completed ☐

13. Refer to the service manual; and locate the resistance checks for the solenoid. What are the specifications?

_____

_____

14. Measure the resistance of the solenoid across the test points specified. What were your results?

_____

_____

15. What can you conclude from these tests?

_____

_____

Problems Encountered

_____

_____

_____

Instructor's Comments

_____

_____

_____

AUTOMATIC TRANSMISSIONS AND TRANSAXLES JOB SHEET 8

Diagnose Control Systems

Name _____ Station _____ Date _____

NATEF Correlation

This Job Sheet addresses the following NATEF tasks:

A.9. Diagnose mechanical and vacuum control system concerns; determine necessary action.

A.13. Diagnose electrical/electronic concerns using principles of electricity (Ohm's Law).

C.1. Inspect, adjust, or replace (as applicable) vacuum modulator; inspect and repair or replace lines and hoses.

Objective

Upon completion of this job sheet, you will be able to diagnose electronic, mechanical, hydraulic, and vacuum control system concerns. You will also be able to inspect, adjust, or replace vacuum modulators; and inspect, repair, or replace vacuum lines and hoses.

Tools and Materials

Component locator   Digital multimeter

Small mirror   Hand-held vacuum pump

Flashlight

Protective Clothing

Goggles or safety glasses with side shields

Describe the vehicle being worked on:

Year _____ Make _____ Model _____

VIN _____ Engine type and size _____

PROCEDURE

1. Check all electrical connections to the transmission. On transaxles, the connectors can normally be inspected through the engine compartment, whereas they can only be seen from under the vehicle on longitudinally mounted transmissions. Make sure they are tight and not damaged. Record your findings.

   _____

   _____

   _____

2. Now, release the locking tabs of the connectors and disconnect them, one at a time, from the transmission. Carefully examine them for signs of corrosion, distortion, moisture, and transmission fluid. A connector or wiring harness may deteriorate if ATF reaches it. Also check the connector at the transmission. Using a small mirror and flashlight may help you get a good look at the inside of the connectors. Record your findings.

_____

_____

3. Inspect the entire transmission wiring harness for tears and other damage. Record your findings.

_____

_____

4. Because the operation of the engine and transmission are integrated through the control computer, a faulty engine sensor or connector may affect the operation of both. The engine control sensors that are the most likely to cause shifting problems are the throttle-position sensor, MAP sensor, and vehicle speed sensor. Locate these sensors and describe their location.

_____

_____

_____

5. Remove the electrical connector from the TP sensor and inspect both ends for signs of corrosion and damage. Poor contact can cause the transmission to miss shifts. Record your findings.

_____

_____

6. Inspect the wiring harness to the TP sensor for evidence of damage. Record your findings.

_____

_____

7. Check both ends of the three-pronged connector and wiring at the MAP sensor for corrosion and damage. Record your findings.

_____

_____

8. Check the condition of the vacuum hose. Record your findings.

_____

_____

9. Check the speed sensor's connections and wiring for signs of damage and corrosion. Record your findings.

_____

_____

10. Record your conclusions from the visual inspection.

_____

_____

_____

_____

11. Move the gear selector lever slowly until it clicks into the PARK position. Turn the ignition key to the START position. If the starter operates, the PARK position is correct. After checking the PARK position, move the lever slowly toward the NEUTRAL position until the lever drops at the end of the N stop in the selector gate. If the starter also operates at this point, neutral is okay. By completing this routine, what two things did you check?

_____

_____

12. If the engine does not start in P or N, the neutral safety switch or the gear selector linkage needs adjustment or repair. Some neutral safety switches are non-adjustable. If these prevent starting in P and/or N and the gear selector linkage is correctly adjusted, they should be replaced. A voltmeter can be used to check the switch for voltage when the ignition key is turned to the START position with the shift lever in P or N. If there is no voltage, the switch should be adjusted or replaced. Record your findings.

_____

_____

13. To adjust a typical neutral safety switch, place the shift lever in Neutral.   Task Completed ☐

14. Loosen the attaching bolts for the switch.   Task Completed ☐

15. Using an aligning pin, move the switch until the pin falls into the hole in its rotor.   Task Completed ☐

16. Tighten the attaching bolts.   Task Completed ☐

17. Recheck the switch for continuity. If no voltage is present, replace the switch.   Task Completed ☐

18. Many older transmissions are fitted with a vacuum modulator. Locate it and describe its location.

_____

_____

19. Diagnosing a vacuum modulator begins with checking the vacuum at the line or hose to the modulator. Record your findings.

_____

_____

20. Check the modulator itself for leaks with a hand-held vacuum pump. The modulator should be able to hold approximately 18 in. Hg. Record your findings.

_____

_____

21. If transmission fluid is found when you disconnect the line at the modulator, the vacuum diaphragm in the modulator is leaking and the modulator should be replaced. Record your findings.

_____

_____

22. If the vacuum source, vacuum lines, and vacuum modulator are in good condition but shift characteristics indicate a vacuum modulator problem, the modulator may need adjustment. To adjust the modulator, prepare to remove it.   Task Completed ☐

23. Loosen the retaining clamp and bolt. (Some units are screwed into the transmission case). While pulling the modulator out of the housing, be careful not to lose the modulator actuating pin, which may fall out as the modulator is removed.   Task Completed ☐

24. Use a hand-held vacuum pump with a vacuum gauge and the recommended gauge pins to adjust the modulator according to specifications. Describe the recommended procedure here.

_____

_____

25. Check the service manual and identify all switches located at the transmission. List these here.

_____

_____

_____

26. Determine the function of each switch. Is it a grounding switch or does it complete or open a power circuit?

_____

_____

_____

27. An ohmmeter can be used to identify the type of switch being used and can be used to test the operation of the switch. There should be continuity when the switch is closed and no continuity when the switch is open. Record your findings.

_____

_____

28. Apply air pressure to the part of the switch that would normally be exposed to oil pressure and check for leaks. Record your findings.

_____

_____

29. Electronically controlled transmissions rely on electrical signals from a speed sensor to control shift timing. The speed sensor used in many late-model transmissions is a permanent magnetic (PM) generator. Locate the speed sensor and describe its location.

_____

_____

30. Raise the vehicle on a lift. Allow the wheels to be suspended and able to rotate freely.   Task Completed ☐

31. Set your Digital multimeter to measure AC voltage.   Task Completed ☐

32. Connect the meter to the speed sensor.   Task Completed ☐

33. Start the engine and put the transmission in gear. Slowly increase the engine's speed until the vehicle is at approximately 20 mph, and then measure the voltage at the speed sensor. Record your findings.

_____

_____

34. Slowly increase the engine's speed and observe the voltmeter. The voltage should increase smoothly and precisely with an increase in speed. Record your findings.

_____

_____

35. A speed sensor can also be tested with it out of the vehicle. Connect an ohmmeter across the sensor's terminals. What was the measured resistance?

_____

_____

36. Locate the specifications for the sensor and compare your readings with specifications.

_____

_____

Problems Encountered

_____

_____

_____

Instructor's Comments

_____

_____

_____

AUTOMATIC TRANSMISSIONS AND TRANSAXLES JOB SHEET 9

Logical Diagnosis

Name _____ Station _____ Date _____

NATEF Correlation

This Job Sheet addresses the following NATEF task:

A.11. Diagnose transmission/transaxle gear reduction/multiplication concerns using driving, driven, and held member (power flow) principles.

Objective

Upon completion of this job sheet, you will be able to identify the most likely sources of a problem by using your understanding of the powerflow of the transmission.

Tools and Materials

A vehicle with an automatic transmission

Service manual

Protective Clothing

Goggles or safety glasses with side shields

Describe the vehicle being worked on:

Year _____ Make _____ Model _____

VIN _____ Engine type and size _____

PROCEDURE

NOTE: *Identifying what planetary gear set members and holding and apply devices are activated in each gear is not merely an exercise to keep you busy. Rather, the understanding of powerflow through a transmission is an extremely valuable tool for diagnostics. This job sheet is designed to make you think about the powerflow and the transmission parts involved while the transmission is operating in a particular gear range. This job sheet is not to be completed on a vehicle, rather it should be completed with you sitting down and thinking and applying your understanding of planetary gear action and the powerflow of a particular transmission to answer these questions.*

1. Describe the transmission in the vehicle being tested (include the number of forward gears, clutches, bands, and electronic controls):

_____

_____

_____

_____

2. When the engine is running and the transmission is in Park, what parts of the transmission are rotating with the engine?

_____

_____

3. When the transmission is in Park, what parts could be the cause of a noise or vibration?

_____

_____

4. If Park does not properly engage and hold, what are the possible problems?

_____

_____

5. When the engine is running and the transmission is in Neutral, what parts of the transmission are rotating with the engine?

_____

_____

6. When the transmission is in Neutral, what parts could be the cause of a noise or vibration?

_____

_____

7. When the engine is running and the transmission is in Drive, what parts of the transmission are rotating with the engine during all forward gear ranges?

_____

_____

8. When the transmission is in Drive, what parts could be the cause of a noise or vibration?

_____

_____

9. When the engine is running and the transmission is in First gear, what parts of the transmission are rotating with the engine?

_____

_____

10. When the transmission is in First gear, what parts could be the cause of a noise or vibration?

_____

_____

11. When the transmission is in First gear, what could be the cause of engagement problems?

_____

_____

12. When the engine is running and the transmission is in Second gear, what parts of the transmission are rotating with the engine?

_____

_____

13. When the transmission is in Second gear, what parts could be the cause of a noise or vibration?

_____

_____

14. When the transmission is in Second gear, what could be the cause of engagement problems?

_____

_____

15. When the engine is running and the transmission is in Third gear, what parts of the transmission are rotating with the engine?

_____

_____

16. When the transmission is in Third gear, what parts could be the cause of a noise or vibration?

_____

_____

17. When the transmission is in Third gear, what could be the cause of engagement problems?

_____

_____

18. When the engine is running and the transmission is in Fourth gear, what parts of the transmission are rotating with the engine?

_____

_____

19. When the transmission is in Fourth gear, what parts could be the cause of a noise or vibration?

_____

_____

20. When the transmission is in Fourth gear, what could be the cause of engagement problems?

   _____

   _____

21. When the engine is running and the transmission is in Fifth gear, what parts of the transmission are rotating with the engine?

   _____

   _____

22. When the transmission is in Fifth gear, what parts could be the cause of a noise or vibration?

   _____

   _____

23. When the transmission is in Fifth gear, what could be the cause of engagement problems?

   _____

   _____

24. When the engine is running and the transmission is in Reverse gear, what parts of the transmission are rotating with the engine?

   _____

   _____

25. When the transmission is in Reverse gear, what parts could be the cause of a noise or vibration?

   _____

   _____

26. When the transmission is in Reverse gear, what could be the cause of engagement problems?

   _____

   _____

Problems Encountered

_____

_____

_____

Instructor's Comments

_____

_____

_____

AUTOMATIC TRANSMISSIONS AND TRANSAXLES JOB SHEET 10

Using Pascal's Law to Help with Diagnostics

Name _____ Station _____ Date _____

NATEF Correlation:

This Job Sheet addresses the following NATEF task:

> **A.12.** Diagnose pressure concerns in the transmission using hydraulic principles (Pascal's Law).

Objective

Upon completion of this job sheet, you will be able to apply the basic principles of Pascal's Law to operation of an automatic transmission and to the results of a pressure test.

Tools and Materials

A good text book

Protective Clothing

None

Describe the vehicle being worked on:

Year _____ Make _____ Model _____

VIN _____ Engine type and size _____

PROCEDURE

1. The hydraulic operation of an automatic transmission is based on the principles of Pascal's Law. Briefly describe how this law applies to the functioning of automatic transmissions.

   _____

   _____

2. Briefly explain why incorrect fluid pressures caused by a bad oil pump would affect all operating modes of a transmission.

   _____

   _____

3. The transmission is fitted with several pressure-regulating and flow-directing valves, briefly describe what they do to the movement of fluid throughout the transmission.

   _____

   _____

4. A servo assembly is used to control the application of a band, which must tightly hold the drum it surrounds when it is applied. The holding capacity of the band is determined by the construction of the band and the pressure applied to it. What does the servo do?

_____

_____

5. If a servo has an area of 10 square inches and has a pressure of 70 psi applied to it, the apply force of the servo will be _____ pounds.

    Task Completed ☐

6. A multiple-disc assembly is also used to stop and hold gear set members. If the fluid pressure applied to the clutch assembly is 70 psi and the diameter of the clutch piston is 5 inches, the force applying the clutch pack is _____ pounds.

    Task Completed ☐

7. List three probable causes for all fluid pressures to be low during a pressure test and explain why.

_____

_____

8. What could possibly be indicated by a great pressure drop between shifts?

_____

_____

9. Why are transmissions checked at various engine speeds and throttle openings?

_____

_____

10. If there is an internal leak in the second gear circuit of the transmission, how would this affect its operation?

_____

_____

11. If the line pressure is too high, how will the operation of the transmission be affected?

_____

_____

Problems Encountered

_____

_____

_____

Instructor's Comments

_____

_____

_____

AUTOMATIC TRANSMISSIONS AND TRANSAXLES JOB SHEET 11

Servicing Linkages

Name _____ Station _____ Date _____

NATEF Correlation

This Job Sheet addresses the following NATEF task:

> **B.1.** Inspect, adjust, or replace throttle (TV) linkages or cables; manual shift linkages or cables; transmission range sensor; check gear select indicator (as applicable).

Objective

Upon completion of this job sheet, you will be able to inspect, adjust, or replace throttle valve and gear selector linkages and cables.

Tools and Materials

Basic hand tools

Protective Clothing

Goggles or safety glasses with side shields

Describe the vehicle being worked on:

Year _____ Make _____ Model _____

VIN _____ Engine type and size _____

PROCEDURE

1. If the gear selector linkage is misadjusted, poor gear engagement, slipping, and excessive wear can result. The gear selector linkage should be adjusted so the manual shift valve detent position in the transmission matches the selector level detent and position indicator. To check the adjustment of the linkage, move the shift lever from the PARK position to the lowest DRIVE gear. Detents should be felt at each of these positions. If the detent cannot be felt in either of these positions, the linkage needs to be adjusted. Describe your findings.

   _____

   _____

2. While moving the shift lever, pay attention to the gear position indicator. Although the indicator will move with an adjustment of the linkage, the pointer may need to be adjusted so that it shows the exact gear after the linkage has been adjusted. Describe your findings.

   _____

   _____

CAUTION: *Always set the parking brake before moving the gear selector through its positions.*

3. To adjust a typical floor-mounted gear selector linkage:
   a. Place the shift lever into the DRIVE position.   Task Completed ☐
   b. Loosen the locknuts and move the shift lever until DRIVE is properly   Task Completed ☐
      aligned and the vehicle is in the "D" range.
   c. Tighten the locknut.   Task Completed ☐

4. To adjust a typical cable-type linkage:
   a. Place the shift lever into the PARK position.   Task Completed ☐
   b. Loosen the clamp bolt on the shift cable bracket.   Task Completed ☐
   c. Make sure the preload adjustment spring engages the fork on the bracket.   Task Completed ☐
   d. By hand, pull the shift lever to the front detent position (PARK), then tight-   Task Completed ☐
      en the clamp bolt. The shift linkage should now be properly adjusted.

5. To adjust a typical rod-type linkage:
   a. Loosen or disconnect the shift rod at the shift lever bracket.   Task Completed ☐
   b. Place the gear selector into PARK and the manual shift valve lever into   Task Completed ☐
      the PARK detent position.
   c. With both levers in position, tighten the clamp on the sliding adjustment   Task Completed ☐
      to maintain their relationship. On the threaded type of linkage adjust-
      ment, lengthen or shorten the connection as needed.

6. On some vehicles, you may need to adjust the neutral safety switch after
   resetting the linkage. Did you need to do this? Why?

   _____

   _____

7. After adjusting any type of shift linkage, recheck it for detents throughout
   its range. Make sure a positive detent is felt when the shift lever is placed
   into the PARK position, as a safety measure. Describe your findings.

   _____

   _____

8. If you are unable to make an adjustment, the levers' grommets may be badly   Task Completed ☐
   worn or damaged and should be replaced. When it is necessary to disas-
   semble the linkage from the levers, the plastic grommets used to retain the
   cable or rod should be replaced. Use a prying tool to force the cable or rod
   from the grommet, then cut out the old grommet. Pliers can be used to
   snap the new grommets into the levers and the cable or rod into the levers.

9. The throttle valve cable connects the movement of the throttle pedal movement to the throttle valve in the transmission's valve body. On some transmissions the throttle linkage may control both the downshift valve and the throttle valve. Others use a vacuum modulator to control the throttle valve and a throttle linkage to control the downshift valve. Late-model transmissions may not have a throttle cable; instead, they rely on electronic sensors and switches to monitor engine load and throttle plate opening. The action of the throttle valve produces throttle pressure. What does this transmission use to relay throttle pressure to the valve body?

_____

_____

_____

Problems Encountered

_____

_____

_____

Instructor's Comments

_____

_____

_____

AUTOMATIC TRANSMISSIONS AND TRANSAXLES JOB SHEET 12

Visual Inspection and Filter and Fluid Change

Name _____ Station _____ Date _____

NATEF Correlation

This Job Sheet addresses the following NATEF task:

B.2. Service transmission; perform visual inspection; replace fluids and filters.

Objective

Upon completion of this job sheet, you will be able to conduct a preliminary inspection of an automatic transmission or transaxle and replace its fluid and filter.

Tools and Materials

Lift   Lint-free rags

Large drain pan   Pound-inch torque wrench

Protective Clothing

Goggles or safety glasses with side shields

Describe the vehicle being worked on:

Year _____ Make _____ Model _____

VIN _____ Engine type and size _____

Transmission type and model _____

PROCEDURE

1. Properly position the vehicle on a lift.   Task Completed ☐

2. Check the transmission housing for damage, cracks, and signs of leaks. Record your findings.

   _____

   _____

3. Check the end of the extension housing for signs of leakage. Record your findings.

   _____

   _____

4. Check all wiring and mechanical linkages for looseness or damage. Record your findings.

   _____

   _____

5. Check the condition of the cooler lines and hoses. Record your findings.

   _____

   _____

6. Check the condition of the fluid. Check its level, color, smell, and feel. Record your findings.

   _____

   _____

7. Remove or move any part of the vehicle that may interfere with the removal of the oil pan. What did you need to move?

   _____

   _____

   _____

8. Position the oil drain pan under the transmission pan.   Task Completed ☐

9. Loosen all the pan bolts and remove all but three at one end. Why do you do this?

   _____

   _____

10. After most of the fluid has drained from the pan, support the pan with one hand. Now, remove the remaining bolts and pour the rest of the transmission fluid into the drain pan.   Task Completed ☐

11. Carefully inspect the oil pan and the residue in it. Record your findings. What is indicated by what was in the pan?

   _____

   _____

   _____

12. Remove the old pan gasket and wipe the pan clean with a clean lint-free rag.   Task Completed ☐

13. Unbolt the fluid filter from the transmission's valve body. Keep the drain pan under the transmission while doing this.   Task Completed ☐

14. Gather the new filter and gaskets. Compare the old with the new. What did you find? If there is a difference, what should you do?

   _____

   _____

15. Install the new filter onto the valve body and tighten the attaching bolts to the proper specifications. What are the specifications?

   _____

16. Lay the new pan gasket over the sealing area of the oil pan. Make sure the holes line up properly.

Task Completed ☐

17. Position the pan onto the transmission.

Task Completed ☐

18. Install the attaching bolts and hand-tighten each.

Task Completed ☐

19. Tighten the bolts to the specified torque. Make sure you stagger the tightening. What are the specifications? What order did you follow when tightening the bolts?

_____

_____

20. Lower the vehicle.

Task Completed ☐

21. Pour a little less than the required amount of the recommended fluid into the transmission through the dipstick tube. How much fluid did you put in and what type of fluid was it?

_____

22. Start the engine. Look under the vehicle and check for leaks.

Task Completed ☐

23. With the parking brake applied and the brake pedal depressed, move the gear selector through the gears.

Task Completed ☐

24. After the engine reaches normal operating temperature, put the transmission into park.

Task Completed ☐

25. Check the fluid level and correct it if necessary.

Task Completed ☐

Problems Encountered

_____

_____

_____

Instructor's Comments

_____

_____

_____

AUTOMATIC TRANSMISSIONS AND TRANSAXLES JOB SHEET 13

Servicing Governors

Name _____ Station _____ Date _____

NATEF Correlation

This Job Sheet addresses the following NATEF task:

C.2. Inspect, repair, and replace governor assembly.

Objective

Upon completion of this job sheet, you will be able to inspect, repair, and replace a governor assembly.

Tools and Materials

Torque wrench

Protective Clothing

Goggles or safety glasses with side shields

Describe the vehicle being worked on:

Year _____ Make _____ Model _____

VIN _____ Engine type and size _____

Transmission type and model _____

PROCEDURE

1. If the pressure tests suggest that there is a governor problem, it should be removed, disassembled, cleaned, and inspected. Some governors are mounted internally and the transmission must be removed to service the governor. Others can be serviced by removing the extension housing or oil pan, or by detaching an external retaining clamp and then removing the unit. How did you remove yours?

   _____

   _____

   _____

2. To disassemble a typical governor, remove the primary governor valve from its bore in the governor housing.   Task Completed ☐

3. Remove the secondary valve retaining pin, secondary valve spring, and valve.   Task Completed ☐

4. Thoroughly clean and dry these parts.   Task Completed ☐

5. Test each valve in its bore in the governor housing; they should move freely in their bores without sticking or binding. Record your findings.

_____

_____

6. Check the valves for any signs of burning or scoring and replace them if necessary. Record your findings.

_____

_____

7. Inspect the springs for a loss of tension and burning marks and replace if necessary. Record your findings.

_____

_____

8. To reassemble the governor, place the spring around the secondary valve and insert them into the secondary valve bore.   Task Completed ☐

9. Insert the retaining pin into the governor housing pin holes.   Task Completed ☐

10. Install the primary valve into the governor housing.   Task Completed ☐

11. If the governor assembly was removed from the governor support and parking gear, be sure to tighten the bolts to specifications with a torque wrench. After assembly, install the governor and torque the bolts to specifications. What are the specifications?

_____

Problems Encountered

_____

_____

_____

Instructor's Comments

_____

_____

_____

AUTOMATIC TRANSMISSIONS AND TRANSAXLES JOB SHEET 14

Inspecting and Servicing Gaskets and Seals

Name _____ Station _____ Date _____

NATEF Correlation

This Job Sheet addresses the following NATEF task:

C.3. Inspect and replace external seals and gaskets.

Objective

Upon completion of this job sheet, you will be able to install common seals and gaskets in a transmission/transaxle.

Tools and Materials

Service manual
Seal removal and installation tools
Clean ATF

Protective Clothing

Goggles or safety glasses with side shields

Describe the vehicle being worked on:

Year _____ Make _____ Model _____

VIN _____ Engine type and size _____

PROCEDURE

Seals

1. Identify the main components of an automatic transmission that have seals that should be replaced during an overhaul and list them here.

   _____

   _____

   _____

   _____

2. Assuming these will be replaced, make sure the replacement seals are of the type that is recommended by the manufacturer of the transmission.

   Task Completed ☐

3. Check each seal, before installing them, in their own bore. They should be slightly smaller or larger (+ or – 3%) than their groove or bore. Do not assume that because a particular seal came with the overhaul kit it is the correct one. What did you find?

_____

_____

4. After lubricating them, keep the seals clean and free of dirt. Before installing seals, clean the shaft and/or bore area.   Task Completed ☐

5. Carefully check the shaft and/or bore area for damage. File or stone away any burrs or bad nicks and polish the surfaces with a fine crocus cloth, then clean the area to remove the metal particles.   Task Completed ☐

6. Lubricate all seals in clean ATF, unless otherwise instructed by the manufacturer. Remember to use only the proper fluids as stated in the appropriate service manual.   Task Completed ☐

7. O-ring and square-cut seals are used to seal non-rotating parts. These seals are commonly used on oil pumps, servos, clutch pistons, speedometer drives, and vacuum modulators. Where will you install this seal?

_____

_____

8. Coat the entire seal with assembly lube.   Task Completed ☐

9. Install the seal while making sure you do not stretch or distort the seal while you work it into its holding groove. After a square-cut seal is installed, double-check it to make sure it is not twisted. The flat surface of the seal should be parallel with the bore.

10. Did you have any problems?

_____

_____

11. Lip seals are used around rotating shafts and apply pistons. The rigid outer diameter provides a mounting point for the seal and is pressed into a bore. What seal will you be replacing?

_____

_____

12. Set the seal in place. Make sure the lip is facing the correct direction.   Task Completed ☐

13. Use the correct driver to install the seal and be careful not to damage the seal during installation.   Task Completed ☐

14. Teflon or metal sealing rings are commonly used to seal servo pistons, oil pump covers, and shafts. They are designed to provide a seal, they may also allow a controlled amount of fluid leakage to lubricate a bushing. What type of Teflon or metal seal will you be installing for this job sheet?

_____

_____

15. Solid sealing rings are commonly made of a Teflon-based material and never reused. To remove them, cut the seal after it has been pried out of its groove. Installing a new ring requires special tools. What tools do you have to do this?

_____

_____

16. What will happen if you install this type seal without the proper tools?

_____

_____

17. Open-end sealing rings fit loosely into a machined groove. This type of ring is typically removed and installed with a pair of snap ring pliers. Do you have the appropriate pliers?

_____

_____

18. Butt-end sealing rings can be removed with a small screwdriver. To install this type of ring, use a pair of snap ring pliers and expand the ring to move it into position. Do you have the appropriate pliers?

_____

_____

19. Locking-end rings may have hooked ends that connect or ends that are cut at an angle. These seals are removed and installed in the same way as butt-end rings. After these rings are installed, make sure the ends are properly positioned and touching. Where do these seals fit into the transmission you are working on?

_____

_____

20. Check all metal sealing rings for proper fit. Insert them in their bore. Do they feel tight?

_____

_____

21. Check the fit of the rings in their shaft groove. If the ring can move laterally in the groove, the groove is worn. What did you find?

_____

_____

22. Check the side clearance of the ring by placing the ring into its groove and measure the clearance between the ring and the groove with a feeler gauge. Typically, the clearance should not exceed 0.003 inches. What did you find?

_____

_____

23. Before installing the seals, look for nicks in the grooves and for evidence of groove taper or stepping. If the grooves are tapered or stepped, the shaft will need to be replaced. If there are burrs or nicks in the grooves, they can be filed away. What did you find?

_____

_____

24. Install the seals. If the seal has some form of locking ends, these should be interlocked prior to trying the seal in its bore.   Task Completed ☐

Gaskets

1. Gaskets are used to seal non-moving components. However, their sealing ability is critical to the durability of a transmission. How many gaskets do you have in your overhaul kit?

_____

_____

2. Make sure all gasket surfaces are clean and flat. Transmission gaskets should not be installed with any type of liquid adhesive or sealant, unless specifically noted by the manufacturer. If any sealer gets into the valve body, severe damage can result. Also, sealant can clog the oil filter. If a gasket is difficult to install, a thin coating of transmission assembly lube can be used to hold the gasket in place.   Task Completed ☐

3. One of the common locations for a gasket is between the main oil pump and the transmission case. Will you be installing a new gasket here?

_____

_____

4. If so, place some petroleum jelly in two or three spots around the oil pump gasket and position it on the housing.   Task Completed ☐

5. Align the pump and install it with care. Tighten the pump attaching bolts to specifications in the specified order. What are those specifications?

_____

_____

6. Another common location for a gasket is the oil pan. Oil pans are typically made of stamped steel and can become distorted around the attaching boltholes. Carefully inspect the sealing surface and describe what you found and what needs to be done to the pan.

Task Completed ☐

Problems Encountered

_____

_____

_____

Instructor's Comments

_____

_____

_____

AUTOMATIC TRANSMISSIONS AND TRANSAXLES JOB SHEET 15

Inspect and Service an Extension Housing

Name _____ Station _____ Date _____

NATEF Correlation

This Job Sheet addresses the following NATEF tasks:

C.4. Inspect extension housing, bushings, and seals; perform necessary action.

C.5. Inspect and replace speedometer drive gear, driven gear, vehicle speed sensor (VSS), and retainers.

Objective

Upon completion of this job sheet, you will be able to inspect and service the extension housing, its bushing and seals and the speedometer drive assembly.

Tools and Materials

Hand file
Puller and driver set
Service manual

Protective Clothing

Goggles or safety glasses with side shields

Describe the vehicle being worked on:

Year _____ Make _____ Model _____

VIN _____ Engine type and size _____

Transmission type and model _____

PROCEDURE

1. With the vehicle on a lift, check for leaks at and around the transmission's extension housing. Record your findings.

   _____

   _____

2. An oil leak stemming from the mating surfaces of the extension housing and the transmission case may be caused by loose bolts. To correct this problem, tighten the bolts to the specified torque. What is the specified torque?

   _____

3. Check the extension housing for cracks, especially around the case mounting surface and the pad that attaches to the transmission mount. Record your findings.

   _____

   _____

4. Also check for signs of leakage at the rear of the extension housing. Fluid leaks from the seal of the extension housing can be corrected with the transmission in the car. Record your findings.

   _____

   _____

5. If the leak was at the rear of the housing, the problem may be the bushing or seal. The clearance between the drive shaft's sliding yoke and the bushing should be minimal. If the clearance is satisfactory, a new oil seal will correct the leak. If the clearance is excessive, a new seal and a new bushing should be installed.

   _____

   _____

6. To begin the procedure for replacing the bushing and seal, remove the drive shaft from the vehicle. What do you need to remember before removing the drive shaft?

   _____

   _____

7. Insert the appropriate puller tool into the extension housing until it grips the front side of the bushing.   Task Completed ☐

8. Pull the seal and bushing from the housing.   Task Completed ☐

9. To replace the bushing, drive a new bushing, with the appropriate driver, into the extension housing. Always make sure this bushing is aligned correctly during replacement, or premature failure can result.   Task Completed ☐

10. Lubricate the lip of the seal, then install the new seal in the extension housing. What did you use to install the seal?

    _____

11. Install the drive shaft.   Task Completed ☐

12. If only the seal needs to be replaced, remove the old seal with a puller.   Task Completed ☐

13. Lubricate the lip of the seal, then install the new seal in the extension housing. What did you use to install the seal?

    _____

14. Install the drive shaft.   Task Completed ☐

15. The vehicle's speedometer can be purely electronic, which requires no mechanical hook-up to the transmission, or it can be driven off the output shaft. An oil leak at the speedometer cable can be corrected by replacing the O-ring seal. Check the seal and record the results.

_____

_____

16. If the seal is bad, remove the speedometer drive assembly from the extension housing by removing the hold-down screw that keeps the retainer in its bore.

Task Completed ☐

17. Carefully pull up on the speedometer cable, pulling the speedometer retainer and drive gear assembly from its bore.

Task Completed ☐

18. While replacing a bad seal, inspect the speedometer drive gear for chips and missing teeth. A damaged drive gear can cause the driven gear to fail; therefore, both should be carefully inspected. On some transmissions the speedometer drive gear is a set of gear teeth machined into the output shaft. Inspect this gear. If the teeth are slightly rough, they can be cleaned up and smoothed with a file. If the gear is severely damaged, the entire output shaft must be replaced. Other transmissions have a drive gear that is splined to the output shaft, held in place by a clip, or driven and retained by a ball that fits into a depression in the shaft. If a clip is used, it should be carefully inspected for cracks or other damage. The drive gear can be removed and replaced, if necessary. Record your findings.

_____

_____

19. Carefully remove the old seal from the speedometer drive.

Task Completed ☐

20. Before replacing the speedometer drive seal, clean the top of the cable's retainer.

Task Completed ☐

21. Inspect the retainer of the driven gear on the speedometer cable. Record your findings.

_____

_____

22. Lightly grease and install the O-ring onto the retainer. Also lubricate the drive gear.

Task Completed ☐

23. Gently tap the retainer and gear assembly into its bore while lining the groove in the retainer with the screw hole in the side of the case.

Task Completed ☐

24. Install the hold-down screw and tighten it in place.

Task Completed ☐

25. If the cause of the leak is not found, prepare to remove the extension housing by removing the driveshaft.

Task Completed ☐

26. Clean the extension housing.

Task Completed ☐

27. Using a known flat surface, check the flatness of the mating surface. Any defects that cannot be removed by light filing indicate that the housing should be replaced. Record your findings.

_____

_____

28. Carefully inspect all bores, whether they are threaded or not. All damaged threaded areas should be repaired. If any condition exists that cannot be adequately repaired, the housing should be replaced. Record your findings.

_____

_____

29. Often the speedometer drive gear is responsible for throwing oil back to the rear bushing. A sheared or otherwise inoperative speedometer gear could cause the extension housing bushing to fail. What is the condition of the speedometer drive gear?

_____

_____

30. Before installing the rear extension housing, assemble the parking pawl pin, washer, spring, and pawl, and any other assembly that is enclosed by the extension housing. Are they assembled properly and in good shape?

_____

_____

31. Install a new extension housing gasket and tighten the housing to the transmission. What are the torque specifications for those bolts?

_____

32. Reinstall the drive shaft.   Task Completed ☐

Problems Encountered

_____

_____

_____

Instructor's Comments

_____

_____

_____

AUTOMATIC TRANSMISSIONS AND TRANSAXLES JOB SHEET 16

Testing Electrical/Electronic Components

Name _____ Station _____ Date _____

NATEF Correlation

This Job Sheet addresses the following NATEF task:

C.6. Diagnose electronic transmission control systems using a scan tool; determine necessary action.

Objective

Upon completion of this job sheet, you will be able to inspect and test, adjust, repair, or replace transmission-related electrical and electronic components.

Tools and Materials

Digital multimeter   Component locator

Test light   Service manual

Self-powered test light   Basic hand tools

Scan tool

Protective Clothing

Goggles or safety glasses with side shields

Describe the vehicle being worked on:

Year _____ Make _____ Model _____

VIN _____ Engine type and size _____

Transmission type and model _____

PROCEDURE

1. Using the service manual and component locator for the vehicle, list all of the electrical and electronic components that are related to the operation of the transmission.

   _____

   _____

   _____

2. Carefully check all electrical wires and connectors for damage, looseness, and corrosion. Record your findings.

   _____

   _____

3. Use an ohmmeter to check the continuity through a connector suspected of being faulty. Record your findings.

_____

_____

4. Check all ground cables and connections. Corroded battery terminals and/or broken or loose ground straps to the frame or engine block will cause problems. Record your findings.

_____

_____

5. Check the fuse or fuses to the control module. To accurately check a fuse, either test it for continuity with an ohmmeter or check each side of the fuse for power when the circuit is activated. Record your findings.

_____

_____

6. Make sure the battery's and the alternator's output voltage is at least 12.6 volts. If a problem is found here, correct it before continuing to diagnose the system. Record your findings.

_____

_____

7. With the system on, measure the voltage dropped across connectors and circuits. Record your findings.

_____

_____

8. Prepare the vehicle for a road test. Have a notebook handy to record events during the test.

    Task Completed ☐

9. Connect the scan tool to the DLC.

    Task Completed ☐

10. Drive the vehicle in a normal manner. All pressure and gear changes should be noted. Also the various computer inputs should be monitored and the readings recorded for future reference. If the scan tool has print capabilities, print out a summary of the road test. If the scanner does not have the ability to give a summary of the road test, you should record this same information after each gear or operating condition change. Summarize the results of the road test.

_____

_____

_____

11. After completing the road test, use the scan tool to retrieve trouble codes. Record your findings.

_____

_____

_____

_____

12. Referring to the service manual, describe what is indicated by the trouble codes.

_____

_____

13. If a problem with shift points was noted during the road test, make sure the transmission is not operating in its default mode. Record your findings.

_____

_____

14. If the self-test sequence pointed to a problem in an input circuit, the input should be tested to determine the exact malfunction. Briefly describe the procedure outlined by the manufacturer.

_____

_____

_____

15. Switches can be tested for operation and for excessive resistance with a voltmeter, test light, or ohmmeter. To check the operation of a switch with a voltmeter or a test light, connect the meter's positive lead to the battery side of the switch. With the negative lead attached to a good ground, the voltage should be measured at this point.

Task Completed ☐

16. Without closing the switch, move the positive lead to the other side of the switch. If the switch is open, no voltage will be present at that point. The amount of voltage present at this side of the switch should equal the amount on the other side when the switch is closed. If the voltage decreases, the switch is causing a voltage drop due to excessive resistance. If no voltage is present on the groundside of the switch with it closed, the switch is not functioning properly and should be replaced. Record your findings.

_____

_____

17. If a switch has been removed from the circuit, it can be tested with an ohmmeter or a self-powered test light. By connecting the leads across the switch connections, the action of the switch should open and close the circuit. Record your findings.

_____

_____

18. The manual lever switch is open or closed, depending on position and can be checked with an ohmmeter. By referring to a wiring diagram, you should be able to determine when the switch should be open. Then connect the meter across the input and output of the switch. Move the lever into the desired position and measure the resistance. An infinite reading is expected when the switch is open. If there is any resistance, the switch should be replaced. Record your findings.

_____

_____

19. The pressure switches used in today's transmissions are either grounding switches or they connect or disconnect two wires. Refer to the wiring diagram to determine the type of switch and test the switch with an ohmmeter. Base your expected results on the type of switch you are testing. By using air pressure you can easily see if the switch works properly or if it has a leak. Record your findings.

_____

_____

20. Another type of switch is a potentiometer. Rather than open and close a circuit, a potentiometer controls the circuit by varying its resistance in response to something. A TP sensor is a potentiometer that sends very low voltage back to the computer when the throttle plates are closed and increases the voltage as the throttle is opened. Most TP sensors receive a reference voltage of 5 volts. What the TP sensor sends back to the computer is determined by the position of the throttle. A TP sensor can be checked with an ohmmeter or a voltmeter. If checked with an ohmmeter, you should be able to watch the resistance across the TP sensor change as the throttle is opened and closed. Often there will be a resistance specification given in the service manual. Compare your reading to the specification. Record your findings.

_____

_____

_____

21. With a voltmeter, you will be able to measure the reference voltage and the output voltage. Both of these should be within specified amounts. If the reference voltage is lower than normal, check the voltage drop across the reference voltage circuit from the computer to the TP sensor. Record your findings.

_____

_____

22. Vehicle speed sensors provide road speed information to the computer. There are basically two types of speed sensors, AC voltage generator sensors and reed style sensors. Both of these rely on magnetic principles. The activity of the sensor can be checked with an ohmmeter. AC voltage generators rely on a stationary magnet and a rotating shaft fitted with iron teeth. Each time a tooth passes through the magnetic field an electrical pulse is present. By counting the number of teeth on the output shaft, you can determine how many pulses per revolution you will measure with a voltmeter set to AC volts. How many teeth are on the output shaft? How many pulses on the meter did you observe?

_____

_____

_____

23. A mass-airflow sensor is used to determine engine load by measuring the mass of the air being taken into the throttle body. The mass airflow sensor is a wire located in the intake air stream that receives a fixed voltage. The wire is designed so that it changes resistance in response to temperature changes. This sensor can be measured with a Digital multimeter set to the Hz frequency range. Check the service manual for specific values. Summarize the specifications for this and record what you observed while checking this sensor.

_____

_____

_____

24. A scan tool can also be used to test a mass-airflow sensor; most scanners have a test mode that monitors mass-airflow sensors. While diagnosing these systems, keep in mind that cold air is denser than warm air. Record your findings.

_____

_____

25. Temperature sensors are designed to change resistance with changes in temperature. A temperature sensor is based on a thermistor. Some thermistors increase resistance with an increase of temperature. Others decrease the resistance as temperature increases. Obviously these sensors can be checked with an ohmmeter. To do so, remove the sensor. Then determine the temperature of the sensor and measure the resistance across it. Compare your reading with the chart of normal resistances given in the service manual. Was the sensor you tested a PTC or NTC?

_____

_____

_____

26. The controls of an electronically controlled transmission direct hydraulic flow through the use of solenoid valves. If you were unable to identify the cause of a transmission problem through the previous checks, you should continue your diagnostics with testing the solenoids. Before continuing, however, you must first determine if the solenoids are case grounded and fed voltage by the computer or if they always have power applied to them and the computer merely supplies the ground. While looking in the service manual to find this, also find the section that tells you which solenoids are on and which are off for each of the different gears. Summarize how the solenoids in your transmission work and how they are switched for the various speed gears and other functions.

_____

_____

_____

_____

27. To begin this test you should secure the tools and/or equipment necessary to manually activate the solenoids. This can be a special tool that allows you to control solenoid activity or a switch panel that does the same thing.   Task Completed ☐

28. To begin this test sequence, disconnect the wiring harness that leads to the solenoids. Connect the switch assembly to the battery.   Task Completed ☐

29. Start the engine and move the shift lever into DRIVE.   Task Completed ☐

30. Through the switch assembly, turn on the solenoid that engages first gear. Record your findings.

_____

_____

_____

_____

31. Increase your speed, then turn on the solenoid for second gear. The transmission should have immediately shifted into second gear. Record your findings.

_____

_____

32. When you want to shift into third, select the combination of solenoid off and on to provide that gear. Record your findings.

_____

_____

33. Do the same for fourth and fifth gears. Record your findings.

_____

_____

34. At times, a solenoid valve will work fine during light throttle operation but will leak when pressure increases. To verify that the valve is leaking, activate the solenoid, then increase engine speed while pulling on the throttle cable. If the solenoid valve leaks, the transmission will downshift. Leaking solenoids should be suspected whenever the transmission shifts roughly under heavy loads or full throttle but shifts fine under light throttle. Record your findings.

_____

_____

35. Solenoids can be checked for circuit resistance and shorts to ground. The test can be conducted at the transmission case connector. By identifying the proper pins in the connector, individual solenoids can be checked with an ohmmeter. Record your findings.

_____

_____

36. Solenoids can also be tested on a bench. Resistance values are typically given in service manuals for each application. Summarize the resistance specifications and the results of your test.

_____

_____

_____

37. A solenoid may be electrically fine but still may fail mechanically or hydraulically. A solenoid's check valve may fail to seat or the porting can be plugged. To check for this, listen to the solenoid when it is activated. When a solenoid affected in this way is activated, it will make a slow dull thud. A good solenoid tends to snap when activated. Record your findings.

_____

_____

Problems Encountered

_____

_____

_____

Instructor's Comments

_____

_____

_____

AUTOMATIC TRANSMISSIONS AND TRANSAXLES JOB SHEET 17

Checking the Transaxle Mounts

Name _____ Station _____ Date _____

NATEF Correlation

This Job Sheet addresses the following NATEF task:

C.7. Inspect, replace, and align power train mounts.

Objective

Upon completion of this job sheet, you will be able to inspect, replace, and align power train mounts.

Tools and Materials

Engine Support fixture

Engine hoist

Basic hand tools

Protective Clothing

Goggles or safety glasses with side shields

Describe the vehicle being worked on:

Year _____ Make _____ Model _____

VIN _____ Engine type and size _____

Transmission type and model _____

PROCEDURE

1. Many shifting and vibration problems can be caused by worn, loose, or broken engine and transmission mounts. Visually inspect the mounts for looseness and cracks. Give a summary of your visual inspection.

   _____

   _____

   _____

2. Pull up and push down on the transaxle case while watching the mount. If the mount's rubber separates from the metal plate or if the case moves up but not down, replace the mount. If there is movement between the metal plate and its attaching point on the frame, tighten the attaching bolts to an appropriate torque. Describe the results of doing this.

   _____

   _____

3. From the driver's seat, apply the foot brake, set the parking brake, and start the engine. Put the transmission into a forward gear and gradually increase the engine speed to about 1500 to 2000 rpm. Watch the torque reaction of the engine on its mounts. If the engine's reaction to the torque appears to be excessive, broken or worn drive train mounts may be the cause. Describe the results of doing this.

_____

_____

4. If it is necessary to replace the transaxle mount, make sure you follow the manufacturer's recommendations for maintaining the alignment of the driveline. Describe the recommended alignment procedures.

_____

_____

_____

_____

5. When removing the transaxle mount, begin by disconnecting the battery's negative cable.   Task Completed ☐

6. Disconnect any electrical connectors that may be located around the mount. Be sure to label any wires you remove to facilitate reassembly.   Task Completed ☐

7. It may be necessary to move some accessories, such as the horn, in order to service the mount without damaging some other assembly.   Task Completed ☐

8. Install the engine support fixture and attach it to an engine hoist.   Task Completed ☐

9. Lift the engine just enough to take the pressure off the mounts.   Task Completed ☐

10. Remove the bolts attaching the transaxle mount to the frame and the mounting bracket, then remove the mount.   Task Completed ☐

11. To install the new mount, position the transaxle mount in its correct location on the frame and tighten its attaching bolts to the proper torque. What is the torque specification?

_____

12. Install the bolts that attach the mount to the transaxle bracket. Before tightening these bolts, check the alignment of the mount.   Task Completed ☐

13. Once you have confirmed that the alignment is correct, tighten all loosened bolts to their specified torque.   Task Completed ☐

14. Remove the engine hoist fixture from the engine, and reinstall all accessories and wires that may have been removed earlier.   Task Completed ☐

Problems Encountered

_____

_____

_____

Instructor's Comments

_____

_____

_____

AUTOMATIC TRANSMISSIONS AND TRANSAXLES JOB SHEET 18

Prepare to Remove a Transaxle or Transmission

Name _____ Station _____ Date _____

NATEF Correlation

This Job Sheet addresses the following NATEF tasks:

D.1.1. Remove and reinstall transmission and torque converter (rear-wheel drive).

D.1.2. Remove and reinstall transaxle and torque converter assembly.

Objective

Upon completion of this job sheet, you will be able to describe the procedures that must be followed in order to remove a transaxle or transmission from a vehicle.

Tools and Materials

Hoist   Drain pan

Transmission jack   Droplight or good flashlight

Engine support fixture   Service manual

Protective Clothing

Goggles or safety glasses with side shields

Describe the vehicle being worked on:

Year _____ Make _____ Model _____

VIN _____ Engine type and size _____

Model and type of transmission _____

PROCEDURE

> **NOTE:** *This job sheet is designed to allow you to take a good look at what would be involved in removing a transmission from an assigned vehicle.*

1. Refer to the service manual and look for any precautions or special instructions that relate to the removal of a transmission from this vehicle. Describe these below:

   _____

   _____

   _____

2. Before beginning to remove the transmission from this vehicle, what is the first thing that should be disconnected? Why?

_____

_____

3. Look under the hood and identify everything that should be removed or disconnected from above before raising the vehicle on the hoist. List these items below:

_____

_____

_____

_____

_____

_____

_____

_____

_____

_____

_____

4. Raise the vehicle to a comfortable working height.   Task Completed ☐

5. Carefully examine the area and identify everything that should be disconnected or removed before unbolting the transmission from the engine. List these items below:

_____

_____

_____

_____

_____

_____

_____

_____

_____

6. Support the transmission/transaxle with the transmission jack.   Task Completed ☐

7. Unbolt the transmission from the engine and slide it away from the engine, making sure the torque converter and oil pump drive is free.

Task Completed ☐

8. Now lower the transmission with the jack and place it in a suitable work area.

Task Completed ☐

9. Normally, reinstallation is the reverse of the removal procedure. Is this true for this vehicle? If not, what special procedures must be followed?

_____

_____

_____

_____

Problems Encountered

_____

_____

_____

Instructor's Comments

_____

_____

_____

AUTOMATIC TRANSMISSIONS AND TRANSAXLES JOB SHEET 19

Disassemble and Inspect a Transmission

Name _____ Station _____ Date _____

NATEF Correlation

This Job Sheet addresses the following NATEF task:

D.1.3. Disassemble, clean, and inspect transmission/transaxle.

Objective

Upon completion of this job sheet, you will be able to disassemble and inspect the major components of a transmission.

Tools and Materials

Compressed air and air nozzle   Lint-free shop towels

Supply of clean solvent   Service manual

Protective Clothing

Goggles or safety glasses with side shields

Describe the vehicle being worked on:

Year _____ Make _____ Model _____

VIN _____ Engine type and size _____

Model and type of transmission _____

PROCEDURE

1. Disassemble the transmission into major units. Set each unit aside until this job sheet refers to it. Describe any problems you encountered while disassembling the transmission. Be sure to follow the procedures given in the service manual while taking the transmission apart.

   _____

   _____

   _____

   _____

2. Clean the planetary gearset in clean solvent and allow it to air dry. If compressed air is used to help the drying process, firmly hold the pinion gears to prevent the bearings from moving.   Task Completed ☐

3. Clean all thrust washers, thrust bearings, and bushings in clean solvent and allow them to air dry.   Task Completed ☐

4. Place one member of the planetary gearset on the sun gear. Rotate the gearset slowly. Do the same for the other parts of the gearset. Describe the feel of the gears' rotation:

_____

_____

_____

_____

5. Inspect each member of the gearset for damaged or worn gear teeth. Describe their condition below:

_____

_____

6. Inspect the splines of each gearset member and describe their condition below:

_____

_____

7. Remove any buildup of material or dirt that may be present between the teeth of the gears.   Task Completed ☐

8. Look at the carrier assemblies and check for cracks or other damage. Describe your findings here.

_____

_____

9. Inspect the entire gearset for signs of discoloration. Describe your findings and explain what is indicated by them.

_____

_____

10. Check the output shaft and its splines for wear, cracking, or other damage. Describe their condition.

_____

_____

11. Carefully inspect the driving shells and drive lugs for wear and damage. Describe their condition.

_____

_____

12. Summarize the condition of the planetary gearsets and shaft.

_____

_____

_____

_____

13. Check the thrust washers and bearings for damage, excessive wear, and distortion. Describe their condition.

_____

_____

14. Check all of the bushings for signs of wear, scoring, and other damage. Describe their condition.

_____

_____

_____

_____

15. Check the shafts that ride in the bushings. Describe their condition.

_____

_____

16. Summarize the condition of the thrust washers, bearings, and their associated shafts.

_____

_____

_____

Problems Encountered

_____

_____

_____

Instructor's Comments

_____

_____

_____

AUTOMATIC TRANSMISSIONS AND TRANSAXLES JOB SHEET 20

Service a Valve Body

Name _____ Station _____ Date _____

NATEF Correlation

This Job Sheet addresses the following NATEF task:

> **D.1.4.** Inspect, measure, clean, and replace valve body (includes surfaces, bores, springs, valves, sleeves, retainers, brackets, check-balls, screens, spacers, and gaskets).

Objective

Upon completion of this job sheet, you will be able to demonstrate the ability to disassemble, clean, inspect, and reassemble a valve body.

Tools and Materials

Transmission stand or holding fixture	Clean solvent
Compressed air and nozzle	Lint-free shop rags
Pound-inch torque wrench	Service manual
Measuring calipers	

Protective Clothing

Goggles or safety glasses with side shields

Describe the vehicle being worked on:

Model and type of transmission: _____

Vehicle the transmission is from: _____

Year _____ Make _____ Model _____

VIN _____ Engine type and size _____

PROCEDURE

1. Mount the transmission to the stand or holding fixture with the valve body facing up.

   Task Completed ☐

2. Remove the valve body attaching screws. Start at the outside bolts and work toward the center if the service manual does not give specific directions.

   Task Completed ☐

3. Remove the valve body assembly and place it in a container of clean solvent.

   Task Completed ☐

4. Remove the end plates and covers from the assembly.

   Task Completed ☐

5. In the space below, draw out a simplified view of the valve body. Note the location of all valves, check balls, and springs.   Task Completed ☐

6. Begin to disassemble the unit. Lay all of the parts on a clean surface in the order in which they were removed. This will help during reassembly. Be careful with the check balls; they may be different sizes. Some transmissions use slightly larger or smaller check balls in circuits. How many check balls were in the valve body? _____ Were they all the same size? _____

7. Remove all of the gaskets from the assembly. Place them aside. Do not throw them away. They will be needed for comparison when installing new gaskets.   Task Completed ☐

8. Thoroughly clean the main body and plates of the assembly. Do not dry the valve body with anything that will leave lint. Allow the unit to air dry or dry it off with a gentle flow of compressed air.   Task Completed ☐

9. Inspect the valve body for damage and cracks. Also check the flatness of the body and the plates. Describe your findings:

_____

_____

_____

10. What are the valves made of? _____

11. Check all the valves for free movement in their bores. Also check the valves for wear, scoring, and signs of sticking. Describe your findings and recommendations:

_____

_____

_____

12. Correctly install the valves and springs in their bores.   Task Completed ☐

13. Replace the end plates or covers. Install the retaining screws by hand, then tighten them to the specific torque. The specified torque is: _____   Task Completed ☐

14. Install the check balls into their proper location.   Task Completed ☐

15. Compare the new gaskets with the old and install the correct ones on the valve body.   Task Completed ☐

16. Align the spring-loaded check balls and all other parts as needed.   Task Completed ☐

17. Place the gasket on top of the transfer plate. Align the gasket holes with the transfer plate and the bores in the valve body.

Task Completed ☐

18. Position the valve body on the transmission. Align the parking and other internal linkages. Then install the retaining screws by hand.

Task Completed ☐

19. Tighten the screws to the specified torque and in the specified order. The specified torque is: _____

Task Completed ☐

Problems Encountered

_____

_____

_____

Instructor's Comments

_____

_____

_____

AUTOMATIC TRANSMISSIONS AND TRANSAXLES JOB SHEET 21

Servo and Accumulator Service

Name _____ Station _____ Date _____

NATEF Correlation

This Job Sheet addresses the following NATEF tasks:

D.1.5. Inspect servo bore, piston, seals, pin, spring, and retainers; determine necessary action.

D.1.6. Inspect accumulator bore, piston, seals, spring, and retainer; determine necessary action.

Objective

Upon completion of this job sheet, you will be able to inspect the servo's and accumulator's bore, piston, seals, pin, spring, and retainers.

Tools and Materials

Snap ring pliers   Small file

Crocus cloth   Knife

Protective Clothing

Goggles or safety glasses with side shields

Describe the vehicle being worked on:

Year _____ Make _____ Model _____

VIN _____ Engine type and size _____

Transmission type and model _____

PROCEDURE

1. On some transmissions, the servo and accumulator assemblies are serviceable with the transmission in the vehicle. Others require the complete disassembly of the transmission. Before disassembling a servo or any other component, carefully inspect the area to determine the exact cause of the leakage. Do this before cleaning the area around the seal. Look at the path of the fluid leakage and identify other possible sources. These sources could be worn gaskets, loose bolts, cracked housings, or loose line connections. Describe your findings.

   _____

   _____

2. Inspect the outside area of the seal. If it is wet, determine if the oil is leaking out or if it is merely a lubricating film of oil. Describe your findings.

   _____

   _____

3. When removing the servo, continue to look for causes of the leak. Check both the inner and outer parts of the seal for wet oil, which means leakage. Describe your findings.

_____

_____

4. When removing the seal, inspect the sealing surface, or lips. Look for unusual wear, warping, cuts and gouges, or particles embedded in the seal. Describe your findings.

_____

_____

5. Remove the retaining rings and pull the assembly from the bore for cleaning.   Task Completed ☐

6. Check the condition of the piston and springs. Cast iron seal rings may not need replacement but rubber and elastomer seals should always be replaced. Describe your findings.

_____

_____

7. Begin the disassembling of an accumulator by removing the accumulator plate snap ring.   Task Completed ☐

8. Remove the accumulator plate, the spring and accumulator pistons. If rubber seal rings are installed on the piston, replace them whenever you are servicing the accumulator.   Task Completed ☐

9. Lubricate the new accumulator piston ring and carefully install it on the piston.   Task Completed ☐

10. Lubricate the accumulator cylinder walls and install the accumulator piston and spring.   Task Completed ☐

11. Install the accumulator plate and retaining snap ring.   Task Completed ☐

12. A servo is serviced in a similar fashion. The servo's piston, spring, piston rod, and guide should be cleaned and dried.   Task Completed ☐

13. Check the servo piston for cracks, burrs, scores, and wear. Servo pistons should be carefully checked for cracks and their fit on the guide pins. Describe your findings.

_____

_____

14. Check the seal groove for defects or damage. Describe your findings.

_____

_____

15. Check cast-iron seal rings to make sure they are able to turn freely in the piston groove. These seal rings are not typically replaced unless they are damaged, so carefully inspect them. Describe your findings.

_____

_____

16. Inspect the servo or accumulator spring for possible cracks. Also check where the spring rests against the case or piston. Describe your findings.

_____

_____

17. Inspect the servo cylinder for scores or other damage. Describe your findings.

_____

_____

18. Move the piston rod through the piston rod guide and check for freedom of movement. Describe your findings.

_____

_____

19. Check the band servo components for wear and scoring. Describe your findings.

_____

_____

20. When reassembling the servo, lubricate the seal ring with ATF and carefully install it on the piston rod.   Task Completed ☐

21. Lubricate and install the piston rod guide with its snap ring into the servo piston.   Task Completed ☐

22. Install the servo piston assembly, return spring, and piston guide into the servo cylinder.   Task Completed ☐

23. Lubricate and install the new lip seal.   Task Completed ☐

Problems Encountered

_____

_____

_____

Instructor's Comments

_____

_____

_____

AUTOMATIC TRANSMISSIONS AND TRANSAXLES JOB SHEET 22

Reassembly of a Transmission/Transaxle

Name _____ Station _____ Date _____

NATEF Correlation

This Job Sheet addresses the following NATEF tasks:

D.1.7. Assemble transmission/transaxle.

D.3.1. Measure endplay or preload; determine necessary action.

Objective

Upon completion of this job sheet, you will be able to properly measure and set endplay and preload during the assembly of a transmission or transaxle.

Tools and Materials

Basic hand tools   Air nozzle

Clean ATF in a tray   Dial indicator

Petroleum jelly

Protective Clothing

Goggles or safety glasses with side shields

Describe the vehicle being worked on:

Year _____ Make _____ Model _____

VIN _____ Engine type and size _____

Transmission type and model _____

PROCEDURE

1. Before proceeding with the final assembly of all components, it is important to verify that the case, housing, and parts are clean and free from dust, dirt, and foreign matter.   Task Completed ☐

2. Coat all parts with the proper type of ATF. Soak bands and clutches in the fluid for at least 15 minutes before installing them. All new seals and rings should have been installed before beginning final assembly.   Task Completed ☐

3. Examine all thrust washers carefully and coat them with petroleum jelly before placing them in the housing.   Task Completed ☐

4. Install the thrust ring, piston return spring, thrust washer, and one-way clutch inner race into the case. Align and start the bolts into the inner race from the rear of the case. Torque the bolts to specifications. What are the specifications?

_____

5. Lubricate and install the rear piston into the case.   Task Completed ☐

6. After determining the correct number of friction and steel plates, install the steel dished plate first, then the steel and friction plates, and finally the retaining plate and snap ring. How many steels did you have?

   _____

7. Using a suitable blowgun with a rubber tapered tip, air check the rear brake operation. What were the results?

   _____

   _____

8. After the rear brake has been completely assembled, measure the clearance between the snap ring and the retainer plate. Select the proper thickness of retaining plate that will give the correct ring to plate clearance if the measurement does not meet the specified limits. What were the results?

   _____

   _____

9. Slide the governor distributor assembly onto the output shaft from the front of the shaft, install the shaft and governor distributor into the case, using care not to damage the distributor rings.   Task Completed ☐

10. On some models, the output shaft, bearing, and appropriate gauging shims are placed into the transmission housing. The output shaft washer and bolt is then installed. While holding the output shaft and gear assembly, torque the output shaft nut to specifications. What are the specifications?

    _____

11. Install a dial indicator and check the travel of the output shaft as it is pushed and pulled. What were the results?

    _____

    _____

12. Remove the gauging shims and install the correct sizes of service shims, output shaft gear, washer, and nut.   Task Completed ☐

13. Torque the output shaft nut to specifications. Using an inch-pound torque wrench, check the turning torque of the output shaft and compare this reading to specifications. What were the results?

    _____

    _____

14. Place the small thrust washer on the pilot end of the transaxle output shaft.   Task Completed ☐

15. Place the rear clutch assembly, front clutch drum, turbine shaft, and thrust washer into the housing.   Task Completed ☐

16. Locate and align the rear clutch over its hub. Gently move the rear clutch and turbine shaft around, rotating the assembly to engage the teeth of the friction discs with the rear clutch hub. Align the direct clutch assembly over the front clutch hub. Move the input shaft back and forth, rotating it so the front clutch friction discs engage with the front clutch hub.

Task Completed ☐

17. Position the thrust washer to the back of the rear planetary carrier.

Task Completed ☐

18. Install the rear planetary carrier and thrust washer into the housing to engage the rear planetary ring gear.

Task Completed ☐

19. Install the front thrust washer and the drive shell assembly, engaging the common sun gear with the planetary pinions in the rear planetary carrier.

Task Completed ☐

20. Assemble the front planetary gear assembly into the front planetary ring gear. Make sure the planetary pinion gear shafts are securely locked to the planetary carrier.

Task Completed ☐

21. Install the one-way sprag into the one-way clutch outer race with the arrow on the sprag facing the front of the transmission.

Task Completed ☐

22. Install the connecting drum with sprag by rotating the drum clockwise using a slight pressure and wobbling to align the plates with the hub and sprag assembly. The connecting drum should now be free to rotate clockwise only. This check will verify that the sprag is correctly installed and operative. What were the results?

_____

_____

23. Install the rear internal gear and the shaft's snap ring.

Task Completed ☐

24. Secure the thrust bearing with petroleum jelly and install the rear planet carrier and the snap ring.

Task Completed ☐

25. Assemble the front and rear clutch drum assemblies together and lay them flat on the bench.

Task Completed ☐

26. Make sure the rear hub thrust bearing is properly seated, then measure from the face of the front clutch drum to the top of the thrust bearing. What did you measure and how does it compare to the specifications?

_____

_____

_____

27. Install the thrust washer and pump front bearing race to the pump.

28. Measure from the pump shaft (bearing race included) to the race of the thrust washer. If the thrust washer is not within the limits, replace it with one of the correct thickness. What were the results?

_____

_____

29. Total endplay should now be checked. Set the transmission case on end, front end up.

Task Completed ☐

30. Make sure the thrust bearings are secure with petroleum jelly. Pick up the complete front clutch assembly and install it into the case. Be sure all parts are seated before proceeding with the measurement. Using a dial indicator or caliper, measure the distance from the rear hub thrust bearing to the case. What were the results?

_____

_____

31. Measure the pump with the front bearing race and gasket installed. The tolerance should fall within specifications. If the difference between the measurements is not within tolerance, select the proper size front bearing race. If it is necessary to change the front bearing race, be sure to change the front clutch thrust washers the same amount. What were the results?

_____

_____

32. Install the brake band servo. Use extreme care not to damage the O-rings. Lubricate around the seals.

Task Completed ☐

33. Install and torque the retainer bolts to specifications. What are the specifications?

_____

34. Loosen the piston stem.

Task Completed ☐

35. Install the brake band strut and finger tighten the band servo piston stem just enough to keep the band and strut snug or from falling out. Do not adjust the band at this time.

Task Completed ☐

36. Air check for proper performance. What were the results?

_____

_____

37. Place some petroleum jelly on two or three spots around the oil pump gasket and position it on the transaxle housing.

Task Completed ☐

38. Align the pump and install the pump with care.

Task Completed ☐

39. Tighten the pump attaching bolts to specifications in the specified order. What are the specifications and the specified order?

_____

40. Check the rotation of the input shaft. If the shaft does not rotate, disassemble the transmission to locate the misplaced thrust washer. What were the results?

_____

_____

41. Install bell housing and torque the retaining bolts to specifications. What are the specifications?

_____

42. Adjust the band, after you check to make sure that the brake band strut is correctly installed.

Task Completed ☐

43. Torque the piston stem to specifications. What are the specifications?

_____

44. Back off two (or the number specified by the manufacturer) full turns and secure with locknut. What are the specifications?

_____

45. Tighten the locknut to specifications. What are the specifications?

_____

46. Before proceeding with the installation of the valve body assembly, it is good practice to perform a final air check of all assembled components. This will ensure that you have not overlooked the tightening of any bolts or damaged any seals during assembly. What are the specifications?

_____

47. Assemble the parking pawl assembly. Place the assembly into its position and install the extension housing with a new gasket, then tighten the attaching bolts to the proper specifications. What are the specifications?

_____

48. On transaxles, the differential assembly should be disassembled, cleaned, inspected, and reassembled. After it has been reassembled, measure its endplay with gauging shims. Then select a shim thick enough to correct the endplay. What are the specifications?

_____

49. After installing the proper shims, measure the differential turning torque with an inch-pound torque wrench. What were your results?

_____

_____

50. Install the valve body. Be sure the manual valve is in alignment with the selector pin. Tighten the valve body attaching bolts to the specified torque. What are the specifications?

_____

51. Before installing the vacuum modulator valve, it is good practice to measure the depth of the hole in which it is inserted. This measurement determines the correct rod length to ensure proper performance. Refer to the service manual to determine the correct rod length based on your measurements.

Task Completed ☐

52. Before installing the kickdown solenoid or other solenoids, check to verify that they are operating properly. Connect the solenoid to a 12-volt source and ground the other terminal. What happened?

_____

_____

53. Install the kickdown switch.   Task Completed ☐

54. Before installing the oil pan, check the alignment and operation of the control lever and parking pawl engagement. Make a final check to be sure all bolts are installed in the valve body.   Task Completed ☐

55. Install the oil pan with a new gasket. Torque the bolts to specifications. What are the specifications?

_____

56. Lubricate the oil pump's lip seal and the converter neck before installing the converter.   Task Completed ☐

57. Install the converter, making sure that the converter is properly meshed with the oil pump drive gear.   Task Completed ☐

58. The transmission is now ready for installation into the vehicle. Use the reverse of the removal procedures. Remember to follow proper fluid filling procedures.   Task Completed ☐

Problems Encountered

_____

_____

_____

Instructor's Comments

_____

_____

_____

AUTOMATIC TRANSMISSIONS AND TRANSAXLES JOB SHEET 23

Transmission Cooler Inspection and Flushing

Name _____ Station _____ Date _____

NATEF Correlation

This Job Sheet addresses the following NATEF task:

D.1.8. Inspect, leak test, and flush cooler, lines, and fittings.

Objective

Upon completion of this job sheet, you will be able to inspect, leak test, flush, and replace transmission cooler, lines, and fittings.

Tools and Materials

Line wrenches	Drain pan
Tubing cutter	Compressed air and air nozzle
Tubing flaring kit	Supply of clean solvent or mineral spirits
ATF pressure tester	Lint-free shop towels
Various lengths of rubber hose	Service manual

Protective Clothing

Goggles or safety glasses with side shields

Describe the vehicle being worked on:

Year _____ Make _____ Model _____

VIN _____ Engine type and size _____

Model and type of transmission _____

PROCEDURE

1. Remove the transmission dipstick and describe the condition, smell, and color of the fluid.

   _____

   _____

2. Open the radiator cap (after the engine is cooled down) and check the coolant for traces of ATF. Also check the cap and gasket for signs of ATF. Record your findings.

   _____

   _____

3. Replace the radiator cap   Task Completed ☐

4. Inspect the metal lines and fittings to and from the transmission cooler. Look for damage and signs of leakage. Describe your findings.

   _____

   _____

5. Summarize the results of your inspection.

   _____

   _____

6. Place a drain pan under the fittings that connect the cooler lines to the radiator.   Task Completed ☐

7. Disconnect both cooler lines from the radiator. Plug or cap the lines from the transmission.   Task Completed ☐

8. Plug or cap one fitting at the radiator.   Task Completed ☐

9. Remove the radiator cap.   Task Completed ☐

10. Hold the compressed air nozzle tightly against the open fitting.   Task Completed ☐

11. Apply air pressure through the fitting (no more than 75 psi).   Task Completed ☐

12. Check the coolant in the radiator for signs of air bubbles or air movement. Describe your findings.

    _____

    _____

13. Unplug the other fitting and the cooler lines.   Task Completed ☐

14. Reconnect the cooler lines to the radiator.   Task Completed ☐

15. Summarize the results of the leak test.

    _____

    _____

    _____

    _____

16. Set the parking brake and start the engine. Allow the engine and transmission to reach normal operating temperature before proceeding. Turn off the engine.   Task Completed ☐

17. Remove the transmission dipstick. Place a funnel in the dipstick tube.   Task Completed ☐

18. Raise the vehicle on a hoist.   Task Completed ☐

19. Disconnect the cooler return line at the point where it enters the transmission case.   Task Completed ☐

20. Attach a rubber hose to the disconnected cooler return line.   Task Completed ☐

21. Lower the vehicle and place the end of the hose into the funnel.   Task Completed ☐

22. Start the engine and run it at about 1000 rpm.   Task Completed ☐

23. Observe the flow of fluid into the funnel. Describe the flow and summarize what this indicates.

_____

_____

_____

_____

24. Turn off the engine.   Task Completed ☐

25. Raise the vehicle and remove the rubber hose from the return line.   Task Completed ☐

26. Reconnect the return line.   Task Completed ☐

27. Lower the vehicle and reinstall the dipstick.   Task Completed ☐

   The remainder of this job sheet covers a procedure for flushing the transmission cooler. Check with your instructor before proceeding.

1. Raise the vehicle on a hoist.   Task Completed ☐

2. Place a drain pan under the engine's radiator (or place it under the external transmission cooler if the vehicle is so equipped).   Task Completed ☐

3. Disconnect both cooler lines at the radiator or cooler.   Task Completed ☐

4. Clean the fittings at the ends of the lines and at the radiator or cooler.   Task Completed ☐

5. Connect a rubber hose to the inlet fitting at the radiator.   Task Completed ☐

6. Place the other end of the hose into the drain pan.   Task Completed ☐

7. Attach a hose funnel to the cooler return fitting. Install a funnel in the other end of this hose.   Task Completed ☐

8. Pour small amounts of mineral spirits into the funnel. Observe the flow and the condition of the fluid moving into the drain pan. What did it look like?

_____

_____

9. If there is little flow, apply some air pressure to the hose connected to the return line.   Task Completed ☐

10. Continue pouring mineral spirits into the funnel until the fluid entering the drain pan is clear.   Task Completed ☐

11. Then disconnect the cooler lines at the transmission and place the drain pan at one end of the lines.   Task Completed ☐

12. Install the rubber hose and funnel to the other end of the lines.   Task Completed ☐

13. Pour mineral spirits through the lines until the flow is clear. Describe your findings.

_____

_____

14. Now pour ATF through the cooler lines to remove the mineral spirits.   Task Completed ☐

15. Remove all hoses and reconnect the cooler lines to the transmission and radiator.   Task Completed ☐

16. Check the transmission's fluid level and correct it if necessary.   Task Completed ☐

Problems Encountered

_____

_____

_____

Instructor's Comments

_____

_____

_____

AUTOMATIC TRANSMISSIONS AND TRANSAXLES JOB SHEET 24

Visual Inspection of a Torque Converter

Name _____ Station _____ Date _____

NATEF Correlation

This Job Sheet addresses the following NATEF task:

D.2.1. Inspect converter flex plate, attaching parts, pilot, pump drive, and seal areas.

Objective

Upon completion of this job sheet, you will be able to conduct a visual inspection of a torque converter.

Tools and Materials

Clean white paper towels   Droplight or good flashlight

Hoist   Service manual

Remote starter switch

Protective Clothing

Goggles or safety glasses with side shields

Describe the vehicle being worked on:

Year _____ Make _____ Model _____

VIN _____ Engine type and size _____

Model and type of transmission _____

PROCEDURE

1. Park the vehicle on a level surface.   Task Completed ☐

2. Describe the condition of the fluid (color, condition, and smell).

   _____

   _____

3. What is indicated by the fluid's condition?

   _____

   _____

   _____

4. Connect a remote starter switch to the vehicle and disable the ignition. Explain how you did this:

   _____

   _____

5. Raise the vehicle on a hoist to a comfortable working height.   Task Completed ☐

6. Put on eye protection.   Task Completed ☐

7. Remove the torque converter access cover or shield.   Task Completed ☐

8. Inspect the converter through the access hole. Use the remote starter switch to observe the entire converter. Describe your findings below:

   _____

   _____

   _____

9. Carefully look for signs of fluid leakage around the torque converter and the access cover or shield. Describe your findings below and state what could be causing the leak.

   _____

   _____

   _____

10. Look at the heads of all torque converter bolts. Do they show signs of contact with other parts?

    _____

    _____

    _____

Problems Encountered

_____

_____

_____

Instructor's Comments

_____

_____

_____

AUTOMATIC TRANSMISSIONS AND TRANSAXLES JOB SHEET 25

Inspection and Testing of a Torque Converter

Name _____ Station _____ Date _____

NATEF Correlation

This Job Sheet addresses the following NATEF task:

D.2.2. Measure torque converter endplay and check for interference; check stator clutch.

Objective

Upon completion of this job sheet, you will be able to measure torque converter endplay and check for interference.

Tools and Materials

Special tools for the torque converter

Dial indicator

Protective Clothing

Goggles or safety glasses with side shields

Describe the vehicle being worked on:

Year _____ Make _____ Model _____

VIN _____ Engine type and size _____

Transmission type and model _____

PROCEDURE

1. Detailed diagnosis of a torque converter takes place with the torque converter removed from the vehicle. Inspect the drive studs or lugs used to attach the converter to the flex-plate. Summarize your findings.

   _____

   _____

2. Check the threads of the lugs and studs. If they are damaged slightly, they can be cleaned up with a thread file, tap, or die set. However, if they are badly damaged, the converter should be replaced. The converter should also be replaced if the studs or lugs are loose or damaged. Carefully check the weld spots that position the studs and lugs on the converter. If the welds are cracked, replace the converter. Summarize your findings.

   _____

   _____

3. Check the hub of the converter for scoring, nicks, excessive wear, burrs, and signs of discoloration. If polishing cannot remove the defects in the hub, the converter should be replaced. Summarize your findings.

_____

_____

4. Noises may be caused by internal converter parts hitting each other or hitting the housing. To check for any interference between the stator and turbine, place the converter, face down, on a bench. Summarize your findings.

_____

_____

5. Then install the oil pump assembly. Make sure the oil pump drive engages with the oil pump.   Task Completed ☐

6. Insert the input shaft into the hub of the turbine.   Task Completed ☐

7. Hold the oil pump and converter stationary, then rotate the turbine shaft in both directions. If the shaft does not move freely and/or makes noise, the converter must be replaced. Summarize your findings.

_____

_____

8. To check for any interference between the stator and the impeller, place the transmission's oil pump on a bench and fit the converter over the stator support splines. Rotate the converter until the hub engages with the oil pump drive. Then hold the pump stationary and rotate the converter in a counterclockwise direction. If the converter does not freely rotate or makes a scraping noise during rotation, the converter must be replaced. Summarize your findings.

_____

_____

9. Check endplay with the proper holding tool and a dial indicator with a holding fixture. Insert the holding tool into the hub of the converter and once bottomed, tighten it in place.   Task Completed ☐

10. This locks the tool into the splines of the turbine while the dial indicator is fixed onto the hub. The amount indicated on the dial indicator, as the tool is lifted up, is the amount of endplay inside the converter. If this amount exceeds specifications, replace the converter. Summarize your findings.

_____

_____

11. If the initial visual inspection suggested that the converter has a leak, special test equipment can be used to determine if the converter is leaking. This equipment uses compressed air to pressurize the converter. Leaks are found in much the same way as tire leaks are, that is, the converter is submerged in water and the trail of air bubbles leads the technician to the source of leakage. This test equipment can only be used to check converters with a drain plug and therefore is somewhat limited in its current applications. However, some manufacturers list a procedure for installing a drain plug or draining the converter in their service manuals. Summarize your findings.

_____

_____

Problems Encountered

_____

_____

_____

Instructor's Comments

_____

_____

_____

AUTOMATIC TRANSMISSIONS AND TRANSAXLES JOB SHEET 26

Servicing an Automatic Transmission Oil Pump

Name _____ Station _____ Date _____

NATEF Correlation

This Job Sheet addresses the following NATEF task:

D.2.3. Inspect, measure, and replace oil pump assembly and components.

Objective

Upon completion of this job sheet, you will be able to inspect, measure, and replace the components of an oil pump assembly.

Tools and Materials

Machinist dye

Feeler gauge

Protective Clothing

Goggles or safety glasses with side shields

Describe the vehicle being worked on:

Year _____ Make _____ Model _____

VIN _____ Engine type and size _____

PROCEDURE

1. With the oil pump removed from the transmission, remove the front pump bearing race, front clutch thrust washer, gasket and O-ring. Inspect the pump bodies, pump shaft, and ring groove areas. Record your findings.

   _____

   _____

   _____

2. Mount the pump on a stand and unbolt and separate the pump bodies.   Task Completed ☐

3. Mark the gears with machinist bluing ink or paint before removing them so that the gears will remain in the same relationship during reassembly. What did you use to mark them and where did you mark them?

   _____

   _____

4. Inspect the gears and all internal surfaces for defects and visible wear. Record your findings.

   _____

   _____

5. With the pump mounted on the stand, use a feeler gauge to measure between the outer gear and the crest in the pump housing crest. What tolerance do the specifications call for? What did you measure?

   _____

   _____

6. Measure between the outer gear teeth and the crescent. What tolerance do the specifications call for? What did you measure?

   _____

   _____

7. Place the pump flat on the bench and with a feeler gauge and straightedge, measure between the gears and the pump cover. What tolerance do the specifications call for? What did you measure?

   _____

   _____

8. Measure the clearance between the C-ring and the ring groove. What tolerance do the specifications call for? What did you measure?

   _____

   _____

9. Using the stand to center the pump, torque the securing bolts to specifications. What is the specified torque?

   _____

   _____

10. Replace all O-rings and gaskets.   Task Completed ☐

Problems Encountered

_____

_____

_____

Instructor's Comments

_____

_____

_____

AUTOMATIC TRANSMISSIONS AND TRANSAXLES JOB SHEET 27

Checking Thrust Washers, Bushings, and Bearings

Name _____ Station _____ Date _____

NATEF Correlation

This Job Sheet addresses the following NATEF tasks:

D.3.2. Inspect, measure, and replace thrust washers and bearings.

D.3.4. Inspect bushings; determine necessary action.

Objective

Upon completion of this job sheet, you will be able to inspect, measure, and replace thrust washers and bearings and inspect the bushings in a transmission/transaxle.

Tools and Materials

Wire-type feeler gauge set

Bushing puller tool

Bushing driver set

Protective Clothing

Goggles or safety glasses with side shields

Describe the vehicle being worked on:

Year _____ Make _____ Model _____

VIN _____ Engine type and size _____

PROCEDURE

1. The best time to inspect thrust washers, bearings, and bushings is during disassembly. The bushings should be inspected for pitting and scoring. Describe their condition.

   _____

   _____

2. Check the depth to which the bushings are installed and the direction of the oil groove, if so equipped, before you remove them. Many bushings that are used in the planetary gearing and output shaft areas have oiling holes in them. Be sure to line the oiling holes up correctly during installation or you may block off oil delivery and destroy the gear train. Describe their condition.

   _____

   _____

3. Observe the lateral movement of the shaft that fits into the bushing. Any noticeable lateral movement indicates wear and the bushing should be replaced. Describe your findings.

_____

_____

4. The amount of clearance between the shaft and the bushing can be checked with a wire-type feeler gauge. Insert the wire between the shaft and the bushing. If the gap is greater than the maximum allowable, the bushing should be replaced. What are the specifications for this gap and how do they compare to your measurement?

_____

_____

5. Measure the inside diameter of the bushing and the outside diameter of the shaft with a vernier type caliper or micrometer. Compare the two and state your conclusions.

_____

_____

6. Most bushings are press-fit into a bore. To remove them they are driven out of the bore with a properly sized bushing tool. Some bushings can be removed with a slide hammer fitted with an expanding or threaded fixture that grips to the inside of the bushing. Another way to remove bushings is to carefully cut one side of the bushing and collapse it. Once collapsed, the bushing can be easily removed with a pair of pliers. What did you use to remove the bushing?

_____

_____

7. Small-bore bushings located in areas where it is difficult to use a bushing tool can be removed by tapping the inside bore of the bushing with threads that match a bolt that fits into the bushing. After the bushing has been tapped, insert the bolt and use a slide hammer to pull the bolt and bushing out of its bore.

Task Completed ☐

8. All new bushings should be pre-lubed during transmission assembly and installed with the proper bushing driver. Make sure they are not damaged and are fully seated in their bore.

Task Completed ☐

9. The purpose of a thrust washer is to support a thrust load and keep parts from rubbing together. Selective thrust washers come in various thicknesses to take up clearances and adjust shaft endplay. Flat thrust washers and bearings should be inspected for scoring, flaking, and wear that goes through to the base material. Describe their condition.

_____

_____

10. Flat thrust washers should also be checked for broken or weak tabs. These tabs are critical for holding the washer in place. On metal type flat thrust washers, the tabs may appear cracked at the bend of the tab; however, this is a normal appearance. Describe their condition.

   _____

   _____

11. Only damaged plastic thrust washers will show wear. The only way to check their wear is to measure the thickness and compare it to a new part. Describe their condition.

   _____

   _____

12. Proper thrust washer thicknesses are important to the operation of an automatic transmission. Always follow the recommended procedure for selecting the proper thrust plate. Use a petroleum jelly type lubricant to hold thrust washers in place during assembly.

    Task Completed ☐

    CAUTION: *Never use white lube or chassis lube. These greases will not mix with the fluid and can plug up orifices, passages, and hold check balls off their seats.*

13. All bearings should be checked for roughness before and after cleaning.

    Task Completed ☐

14. Carefully examine the inner and outer races, and the rollers, needles, or balls for cracks, pitting, etching, or signs of overheating. Describe their condition.

   _____

   _____

15. Give a summary of your inspection.

   _____

   _____

   _____

Problems Encountered

_____

_____

_____

Instructor's Comments

_____

_____

_____

_____

AUTOMATIC TRANSMISSIONS AND TRANSAXLES JOB SHEET 28

Servicing Oil Delivery Seals

Name _____ Station _____ Date _____

NATEF Correlation

This Job Sheet addresses the following NATEF task:

D.3.3. Inspect oil delivery seal rings, ring grooves, and sealing surface areas.

Objective

Upon completion of this job sheet, you will be able to inspect oil delivery seal rings, ring grooves, and sealing surface areas.

Tools and Materials

Petroleum jelly   Crocus cloth

Seal driver tools   Feeler gauge set

Protective Clothing

Goggles or safety glasses with side shields

Describe the vehicle being worked on:

Year _____ Make _____ Model _____

VIN _____ Engine type and size _____

Transmission type and model _____

PROCEDURE GUIDELINES

- Three types of seals are used in automatic transmissions: O-ring and square-cut (lathe-cut), lip, and sealing rings. These seals are designed to stop fluid from leaking out of the transmission and to stop fluid from moving into another circuit of the hydraulic circuit.

- O-ring and square-cut seals are used to seal non-rotating parts. When installing a new O-ring or square-cut seal, coat the entire surface of the seal with assembly lube or petroleum jelly. Make sure you don't stretch or distort the seal while you work it into its holding groove. After a square-cut seal is installed, double check it to make sure it is not twisted. The flat surface of the seal should be parallel with the bore. If it is not, fluid will easily leak past the seal.

- Lip seals that are used to seal a shaft typically have a metal flange around their outside diameter. The shaft rides on the lip seal at the inside diameter of the seal assembly. The rigid outer diameter provides a mounting point for the lip seal and is pressed into a bore. Once pressed into the bore, the outer diameter of the seal prevents fluid from leaking into the bore, while the inner lip seal prevents leakage past the shaft.

- Piston lip seals are set into a machined groove on the piston. This type of lip seal is not housed in a rigid metal flange. They are designed to be flexible and provide a seal while the piston moves up and down. While the piston moves, the lip flexes up and down. The most important thing to keep in mind while installing a lip seal is: make sure the lip is facing in the correct direction. The lip should always be aimed toward the source of pressurized fluid. If installed backwards, fluid under pressure will easily leak past the seal. Also remember to make sure the surfaces to be sealed are clean and not damaged.

- Teflon or metal sealing rings are commonly used to seal servo pistons, oil pump covers, and shafts. These rings may be designed to provide for a seal, but they may also be designed to allow a controlled amount of fluid leakage. Sealing rings are either solid rings or are cut. Cut sealing rings are one of three designs: open-end, butt-end, or locking-end.

- Solid sealing rings are made of a Teflon-based material and are never reused. To remove them, simply (but carefully) cut the seal after it has been pried out of its groove. Installing a new solid sealing ring requires special tools. These tools allow you to stretch the seal while pushing it into position. Never attempt to install a solid seal without the proper tools. Because these seals are soft, they are easily distorted and damaged.

- Open-end sealing rings fit loosely into a machined groove. The ends of the rings do not touch when they are installed. This type of ring is typically removed and installed with a pair of snap ring pliers. The ring should be expanded just enough to move it off or onto the shaft.

- Butt-end sealing rings are designed so that their ends butt up or touch each other once the seal is in place. This type of seal can be removed with a small screwdriver. The blade of the screwdriver is used to work the ring out of its groove. To install this type of ring, use a pair of snap ring pliers and expand the ring to move it into position.

- Locking-end rings may have hooked ends that connect or have ends that are cut at an angle to hold the ends together. These seals are removed and installed in the same way as butt-end rings. After these rings are installed, make sure the ends are properly positioned and touching.

- All seals should be checked in their own bores prior to installation. They should be slightly smaller or larger (+ or − 3%) than their groove or bore. If a seal is not the proper size, find one that is. Do not assume that because a particular seal came with the overhaul kit it is the correct one.

- Never install a seal when it is dry. The seal should slide into position and allow the part it seals to slide into it. A dry seal is easily damaged during installation.

- Install only genuine seals recommended by the manufacturer of the transmission.

PROCEDURE

1. Before installing seals, clean the shaft and/or bore area.   Task Completed ☐

2. Carefully inspect these areas for damage. File or stone away any burrs or bad nicks and polish the surfaces with a fine crocus cloth, then clean the area to remove the metal particles. Describe your findings.

_____

_____

3. Lubricate the seal, especially any lip seals, to ease installation.   Task Completed ☐

4. All metal sealing rings should also be checked for proper fit. Since these rings seal on their outer diameter, the seal should be inserted in its bore and should feel tight there. If the seal has some form of locking ends, these should be interlocked prior to trying the seal in its bore. Describe your findings.

_____

_____

5. Check the fit of the sealing rings in their shaft groove. Describe your findings.

_____

_____

6. Check the side clearance of the ring, by placing the ring into its groove and measuring the clearance between the ring and the groove with a feeler gauge. Describe your findings.

_____

_____

7. While checking the clearance, look for nicks in the grooves and for evidence of groove taper or stepping. Describe your findings.

_____

_____

8. Use the correct driver when installing a seal and be careful not to damage the seal during installation.

Task Completed ☐

Problems Encountered

_____

_____

_____

Instructor's Comments

_____

_____

_____

AUTOMATIC TRANSMISSIONS AND TRANSAXLES JOB SHEET 29

Servicing Planetary Gear Assemblies

Name _____ Station _____ Date _____

NATEF Correlation

This Job Sheet addresses the following NATEF task:

D.3.5. Inspect and measure planetary gear assembly (includes sun gear, ring gear, thrust washers, planetary gears, and carrier assembly); determine necessary action.

Objective

Upon completion of this job sheet, you will be able to inspect and measure planetary gear assemblies.

Tools and Materials

Snap ring pliers

Feeler gauge set

Protective Clothing

Goggles or safety glasses with side shields

Describe the vehicle being worked on:

Year _____ Make _____ Model _____

VIN _____ Engine type and size _____

Transaxle type and model _____

PROCEDURE

1. The planetary gears used in automatic transmissions are the helical type gear and all gear teeth should be inspected for chips or stripped teeth. Describe their general condition before disassembling the gear set.

   _____

   _____

2. Any gear that is mounted to a splined shaft needs the splines checked for mutilation or shifted splines. Record your findings.

   _____

   _____

3. Note any discoloration of the parts and explain the cause for it.

   _____

   _____

4. Check the planetary pinion gears for loose bearings. Record your findings.

_____

_____

5. Check each gear individually by rolling it on its shaft to feel for roughness or binding of the needle bearings. Wiggle the gear to be sure it is not loose on the shaft. Looseness will cause the gear to whine when it is loaded. Record your findings.

_____

_____

6. Inspect the gears' teeth for chips or imperfections as these will also cause whine. Record your findings.

_____

_____

7. Check the gear teeth around the inside of the front planetary ring gear. Record your findings.

_____

_____

8. Check the fit between the front planetary carrier to the output shaft splines. Record your findings.

_____

_____

9. Remove the snap ring and thrust washer from the front planetary ring gear. Record your findings.

_____

_____

10. Examine the thrust washer and the outer splines of the front drum for burrs and distortion. Record your findings.

_____

_____

11. With the snap ring removed, the front planetary carrier can be removed from the ring gear. Check the planetary carrier gears for endplay by placing a feeler gauge between the planetary carrier and the planetary pinion gear. Compare the endplay to specifications. Record your findings.

_____

_____

12. Check the splines of the sun gear. Record your findings.

_____

_____

13. Sun gears should have their inner bushings inspected for looseness on their respective shafts. Record your findings.

_____

_____

14. Check the fit of the sun shell to the sun gear and inspect the shell for cracks especially at the point where the gears mate with the shell. Record your findings.

_____

_____

15. Check the sun shell for a bell-mouthed condition where it is tabbed to the clutch drum. Any variation from a true round should be considered junk and should not be used. Record your findings.

_____

_____

16. Look at the tabs and check for the best fit into the clutch drum slots. This involves trial fitting the shell and drum at all the possible combinations and marking the point where they fit the tightest. Record your findings.

_____

_____

17. Check the gear carrier for cracks and other defects. Record your findings.

_____

_____

18. Check the thrust bearings for excessive wear and, if required, correct the input shaft thrust clearance by using a washer with the correct thickness. Record your findings.

_____

_____

19. To determine the correct thickness, measure the thickness of the existing thrust washer and compare it to the measured endplay. All the pinions should have about the same endplay. Record your findings.

_____

_____

20. Replace all defective parts and reassemble the gear set. Task Completed ☐

Problems Encountered

_____

_____

_____

Instructor's Comments

_____

_____

_____

AUTOMATIC TRANSMISSIONS AND TRANSAXLES JOB SHEET 30

Transmission Case Service

Name _____ Station _____ Date _____

NATEF Correlation

This Job Sheet addresses the following NATEF task:

D.3.6. Inspect case bores, passages, bushings, vents, and mating surfaces; determine necessary action.

Objective

Upon completion of this job sheet, you will be able to inspect case bores, passages, vents, bushings, and mating surfaces.

Tools and Materials

Air nozzle   Straightedge

Crocus cloth   Feeler gauge set

Protective Clothing

Goggles or safety glasses with side shields

Describe the vehicle being worked on:

Year _____ Make _____ Model _____

VIN _____ Engine type and size _____

Transmission type and model _____

PROCEDURE

1. The transmission case should be thoroughly cleaned and all passages blown out.   Task Completed ☐

2. The passages can be checked for restrictions by applying compressed air to each one. If air flows from the other end, there is no restriction. Describe your findings.

   _____

   _____

3. To check for leaks, plug off one end of the passage and apply air to the other. If pressure builds up in that passage, there are probably no leaks in it. Describe your findings.

   _____

   _____

4. Check the fit of the servo piston in the bore without the seal to be sure it has free travel. There should be no tight spots or binding over the whole range of travel. Any deep scratches or gouges that cause binding of the piston will require case replacement. Describe your findings.

_____

_____

5. Accumulator bores are checked the same as servo bores. Describe your findings.

_____

_____

6. Check the oil pump bore at the front of the case. Describe your findings.

_____

_____

7. Case mounted hydraulic clutch bores are prone to the same problems as servo bores. Look for any scratches or gouges in the sealing area that would affect the rubber seals. It is possible to damage these areas during disassembly, so be careful with tools used during overhaul. Describe your findings.

_____

_____

8. Sealing surfaces of the case should be inspected for surface roughness, nicks, or scratches where the seals ride. Imperfections in steel or cast iron parts can usually be polished out with crocus cloth. Describe your findings.

_____

_____

9. Check the passages in the case for cross-tracking of one circuit to another. Fill the circuit with solvent and watch to see if the solvent disappears or leaks away. If the solvent goes down, you should check each part of the circuit to find where the leak is. Describe your findings.

_____

_____

10. Make sure all necessary check balls were in position during disassembly.   Task Completed ☐

11. Check the valve body mounting area for warpage with a straightedge and feeler gauge. This should be done in several locations. If there is a slight burr or high spot, it can be removed by flat filing the surface. Describe your findings.

_____

_____

12. A long straightedge should be laid across the lower flange of the case to check for distortion. Any warpage found here may result in circuit leakage, causing any number of hydraulically related problems. Describe your findings.

_____

_____

13. Check all bellhousing bolt holes and dowel pins. Cracks around the bolt holes indicate that the case bolts were tightened with the case out of alignment with the engine block. Describe your findings.

_____

_____

14. Check all of the bolts that were removed during disassembly for aluminum on the threads. If so, the thread bore is damaged and should be repaired. Thread repair entails the installation of a thread insert or by re-tapping the bore. After the threads have been repaired, make sure you thoroughly clean the case. Describe your findings.

_____

_____

15. The small screens found during teardown should be inspected for foreign material. Describe your findings.

_____

_____

16. Most screens can be removed easily. Care should be taken when cleaning because some cleaning solvents will destroy the plastic screens. Low air pressure (approximately 30 psi) can be used to blow the screens out in a reverse direction.

Task Completed ☐

17. Bushings in a transmission case are normally found in the rear of the case and require the same inspection and replacement techniques as other bushings in the transmission. Always be sure that the oil passage to a pressure fed bushing or bearing is open and free of dirt and foreign material. Describe your findings.

_____

_____

18. Vents are located in the pump body or transmission case and provide for equalization of pressures in the transmission. These vents can be checked by blowing low pressure air through them, squirting solvent or brake cleaning spray through them, or by pushing a small diameter wire through the vent passage. Describe your findings.

_____

_____

Problems Encountered

_____

_____

_____

Instructor's Comments

_____

_____

_____

AUTOMATIC TRANSMISSIONS AND TRANSAXLES JOB SHEET 31

Servicing Internal Transaxle Drives

Name _____ Station _____ Date _____

NATEF Correlation

This Job Sheet addresses the following NATEF task:

D.3.7. Inspect transaxle drive, link chains, sprockets, gears, bearings, and bushings; perform necessary action.

Objective

Upon completion of this job sheet, you will be able to inspect transaxle drive, link chains, sprockets, gears, bearings, and bushings.

Tools and Materials

Marking tool

Machinist rule

Protective Clothing

Goggles or safety glasses with side shields

Describe the vehicle being worked on:

Year _____ Make _____ Model _____

VIN _____ Engine type and size _____

PROCEDURE

1. This inspection is done with the transaxle on a bench and partially disassembled and should be repeated as a double check during reassembly. Begin by checking chain deflection between the centers of the two sprockets. Deflect the chain inward on one side until it is tight.

   Task Completed ☐

2. Mark the housing at the point of maximum deflection.

   Task Completed ☐

3. Then deflect the chain outward on the same side until it is tight.

   Task Completed ☐

4. Again mark the housing in line with the outside edge of the chain at the point of maximum deflection.

   Task Completed ☐

5. Measure the distance between the two marks. If this distance exceeds specifications, replace the drive chain. Describe your findings.

   _____

   _____

6. Be sure to check for an identification mark on the chain during disassembly. These can be painted or dark colored links, and may indicate either the top or the bottom of the chain, so be sure you remember which side was up. How was your chain marked?

   _____

   _____

7. The sprockets should be inspected for tooth wear and for wear at the point where they ride. If the chain was found to be too slack, it may have worn the sprockets in the same manner that engine timing gears wear when the timing chain stretches. A slightly polished appearance on the face of the gears is normal. Describe your findings.

   _____

   _____

8. Check the bearings and bushings used on the sprockets for damage. Describe your findings.

   _____

   _____

9. The radial needle thrust bearings must be checked for any deterioration of the needles and cage. Describe your findings.

   _____

   _____

10. The running surface in the sprocket must also be checked because the needles may pound into the gear's surface during abusive operation. Describe your findings.

    _____

    _____

11. The bushings should be checked for any signs of scoring, flaking, or wear. Describe your findings.

    _____

    _____

12. Based on the above, what parts need to be replaced?

    _____

    _____

    _____

    _____

Problems Encountered

_____

_____

_____

Instructor's Comments

_____

_____

_____

AUTOMATIC TRANSMISSIONS AND TRANSAXLES JOB SHEET 32

Servicing Final Drive Components

Name _____ Station _____ Date _____

NATEF Correlation

This Job Sheet addresses the following NATEF task:

D.3.8. Inspect, measure, repair, adjust, or replace transaxle final drive components.

Objective

Upon completion of this job sheet, you will be able to inspect, measure, and adjust transaxle final drive units.

Tools and Materials

Basic hand tools

Protective Clothing

Goggles or safety glasses with side shields

Describe the vehicle being worked on:

Year _____ Make _____ Model _____

VIN _____ Engine type and size _____

Transaxle type and model _____

PROCEDURE

1. Final drive units may be helical gear or planetary gear units. A careful inspection of the assembly is done with the transaxle disassembled. The helical type should be checked for worn or chipped teeth, overloaded tapered roller bearings, and excessive differential side gear and spider gear wear. Describe your findings.

   _____

   _____

2. Measure the clearance between the side gears and the differential case. Compare your measurement to specifications. Describe your findings.

   _____

   _____

3. Check the fit of the spider gears on the spider gear shaft. Describe your findings.

   _____

   _____

4. Check the assembly's endplay. How do your measurements compare to specifications?

   _____

   _____

5. What is used to preload the side bearings on this transaxle?

   _____

   _____

6. With a torque wrench, measure the amount of rotating torque. Compare your readings against specifications. Describe your findings.

   _____

   _____

7. If the bearing preload and endplay is fine, as is the condition of the bearings, the parts can be reused. However, always install new seals during assembly.

   Task Completed ☐

8. Planetary-type final drives are checked for the same problems as helical types. Check for worn or chipped teeth, overloaded tapered roller bearings, and excessive differential side gear and spider gear wear. Describe your findings.

   _____

   _____

9. The planetary pinion gears need to be checked for looseness or roughness on their shafts and for endplay. Describe your findings.

   _____

   _____

10. Check the endplay of the assembly. How did you do this and how do your measurements compare to specifications?

    _____

    _____

11. What is used to preload the side bearings on this transaxle?

    _____

    _____

12. With a torque wrench, measure the amount of rotating torque. Compare your readings against specifications. Describe your findings.

    _____

    _____

13. What are your conclusions about the final drive unit?

_____

_____

_____

_____

Problems Encountered

_____

_____

_____

Instructor's Comments

_____

_____

_____

AUTOMATIC TRANSMISSIONS AND TRANSAXLES JOB SHEET 33

Servicing the Parking Pawl Assembly

Name _____ Station _____ Date _____

NATEF Correlation

This Job Sheet addresses the following NATEF task:

D.3.9. Inspect and reinstall parking pawl, shaft, spring, and retainer; determine necessary action.

Objective

Upon completion of this job sheet, you will be able to inspect and reinstall the parking pawl, shaft, spring, and retainer.

Tools and Materials

Basic hand tools

Protective Clothing

Goggles or safety glasses with side shields

Describe the vehicle being worked on:

Year _____ Make _____ Model _____

VIN _____ Engine type and size _____

PROCEDURE

1. The parking pawl assembly is typically not hydraulically activated; instead, the gearshift linkage moves the pawl into position to lock the output shaft of the transmission. The parking pawl can be inspected after the transmission is disassembled or, on some transmissions, while the transmission is still in the vehicle. Are you looking at the parking pawl assembly while the transmission is on a bench or in a vehicle?

_____

2. Check the pawl assembly for excessive wear and other damage. Describe your findings.

_____

_____

3. Check to see how firmly the pawl is in place when the gear selector is shifted into the PARK mode. If the pawl can be easily moved out, it should be repaired or replaced. Describe your findings.

_____

_____

4. Examine the engagement lug on the pawl. Make sure it is not rounded off. Describe your findings.

_____

_____

5. Most parking pawls pivot on a pin. This also needs to be checked to make sure there is no excessive looseness at this point. Describe your findings.

_____

_____

6. The spring that pulls the pawl away from the parking gear must also be checked to make sure it can hold the pawl firmly in place. Describe your findings.

_____

_____

7. Check the position and seating of the spring to make sure it will remain in that position during operation. Describe your findings.

_____

_____

8. The push rod or operating shaft must provide the correct amount of travel to engage the pawl to the gear. Make sure the shaft is not bent or that the pivot holes in the internal shift linkage are not worn oblong. Describe your findings.

_____

_____

Problems Encountered

_____

_____

_____

Instructor's Comments

_____

_____

_____

AUTOMATIC TRANSMISSIONS AND TRANSAXLES JOB SHEET 34

Inspecting Apply Devices

Name _____ Station _____ Date _____

NATEF Correlation

This Job Sheet addresses the following NATEF tasks:

D.4.1. Inspect clutch drum, piston, check-balls, springs, retainers, seals, and friction and pressure plates; determine necessary action.

D.4.4. Inspect roller and sprag clutch, races, rollers, sprags, springs, cages, and retainers; determine necessary action.

Objective

Upon completion of this job sheet, you will be able to inspect various apply devices of a transmission.

Tools and Materials

Compressed air and air nozzle   Lint-free shop towels

Supply of clean solvent   Service manual

Protective Clothing

Goggles or safety glasses with side shields

Describe the vehicle being worked on:

Year _____ Make _____ Model _____

VIN _____ Engine type and size _____

Model and type of transmission _____

PROCEDURE

1. Disassemble the transmission into major units. Set each unit aside until this job sheet refers to it. Describe any problems you encountered while disassembling the transmission. Be sure to follow the procedures given in the appropriate service manual while taking the transmission apart.

_____

_____

_____

_____

_____

2. Clean each overrunning clutch assembly in fresh solvent. Allow them to air dry.   Task Completed ☐

3 Check the rollers and sprags for signs of wear or damage. Describe your findings.

_____

_____

4. Check the springs for distortion, distress, and damage. Describe your findings.

_____

_____

5. Check the inner race and the cam surfaces for scoring and other damage. Describe your findings.

_____

_____

6. Check the condition of the snap rings. Describe your findings.

_____

_____

7. Summarize the condition of the overrunning clutch units.

_____

_____

_____

_____

8. Wipe the transmission's bands clean with a dry, lint-free cloth.   Task Completed ☐

9. Check the bands for damage, wear, distortion, and lining faults. Describe your findings.

_____

_____

10. Inspect the band apply struts for damage, distortion, and other damage. Describe your findings.

_____

_____

11. Summarize the conditions of the bands.

_____

_____

_____

_____

12. Clean servo and accumulator parts in fresh solvent and allow them to air dry.   Task Completed ☐

13. Inspect their bores for scoring and other damage. Describe their condition.

_____

_____

14. Check the piston and piston rod for wear, nicks, burrs, and scoring. Describe your findings.

_____

_____

15. Inspect all springs for damage and distortion. Describe their condition.

_____

_____

16. Check the mating surfaces between the piston and the walls of their bores for scoring, wear, nicks, and other damage. Describe your findings.

_____

_____

17. Check the movement of each piston in its bore. Describe that movement.

_____

_____

18. Check the fluid passages for restrictions and clean out any dirt present in the passages. Describe your findings.

_____

_____

19. Summarize the condition of the servos and accumulators in this transmission.

_____

_____

Problems Encountered

_____

_____

_____

Instructor's Comments

_____

_____

_____

AUTOMATIC TRANSMISSIONS AND TRANSAXLES JOB SHEET 35

Checking and Overhauling a Multiple Friction Disc Assembly

Name _____ Station _____ Date _____

NATEF Correlation

This Job Sheet addresses the following NATEF tasks:

D.4.2. Measure clutch pack clearance; determine necessary action.

D.4.3. Air test the operation of clutch and servo assemblies.

Objective

Upon completion of this job sheet, you will have demonstrated the ability to air test clutch packs and servos; and disassemble, inspect, and reassemble a multiple friction disc assembly.

Tools and Materials

Feeler gauge set

Special tools for the assigned transmission

Emery cloth

Crocus cloth

Clean solvent

Lint-free shop towels

Air nozzle

Air test plate

Service manual

Protective Clothing

Goggles or safety glasses with side shields

Describe the transmission being worked on:

Model and type of transmission: _____

Vehicle the transmission is from:

Year _____ Make _____ Model _____

VIN _____ Engine type and size _____

PROCEDURE

1. With the transmission disassembled, set aside each clutch pack for inspection.

   Task Completed ☐

2. Disassemble each clutch pack. Keep each assembly separate from the others. Record the number of steel and friction plates in each assembly.

   _____

   _____

   _____

3. Check the steel plates for discoloration, scoring, distortion, and other damage. Describe their condition:

_____

_____

_____

4. If the steel plates are in good condition, rough up the shiny surface with emery (80-grit) cloth.

Task Completed ☐

5. Inspect the friction plates for wear, damage, and distortion. Describe their condition.

_____

_____

_____

6. Soak all new and reusable friction plates in ATF for at least 15 minutes before reassembling the clutch pack.

Task Completed ☐

7. Check the pressure plate for discoloration, scoring, and distortion. Describe its condition:

_____

_____

_____

8. Inspect the coil springs for distortion and damage. Describe their condition:

_____

_____

_____

9. Inspect the Belleville spring for wear and distortion. Make sure to check the inner fingers. Describe your findings:

_____

_____

_____

10. Check the clutch drums for damaged lugs, grooves, and splines. Also check them for score marks. Describe their condition:

_____

_____

_____

11. Check the movement of the check balls in the drums. Describe your findings:

_____

_____

_____

12. Summarize the condition of the clutch packs:

_____

_____

_____

13. Reassemble the clutch packs with the required new parts and with new seals. Follow the recommended procedure for doing this.

Task Completed ☐

14. Insert a feeler gauge between the pressure plate and the snap ring. What is the measured clearance? _____

15. What is the specified clearance? _____

16. If the measured clearance is not the same as the specified clearance, what should you do?

_____

_____

_____

17. Do what is necessary to correct the clearance.

Task Completed ☐

18. Tip the clutch assembly on its side and insert a feeler gauge between the pressure plate and the adjacent friction plate. What is the measured clearance? _____

19. What is the specified clearance? _____

20. If the measured clearance is not the same as what was specified, what should you do?

_____

_____

_____

21. Do what is necessary to correct the clearance.

Task Completed ☐

22. After the clearance is set, check the service manual for the proper procedure for air testing the clutch pack. Summarize the procedure.

_____

_____

_____

23. Place the pack or the partially assembled transmission in a vise or on a stand.   Task Completed ☐

24. Apply low air pressure to the designated test port. Pay attention to the sound of the air leaking and the activation of the disc pack. Describe what you heard.

   _____

   _____

25. Release the air and listen for the release of the pack. Describe what happened.

   _____

   _____

   _____

26. Based on the above, what are your conclusions?

   _____

   _____

   _____

27. When the transmission is assembled, install the air test plate to the designated area on the transmission. What does the plate attach to?

   _____

   _____

28. What components can be checked with the test plate?

   _____

   _____

   _____

29. Apply low pressure to each of the test ports and record the results.

   _____

   _____

   _____

30. What is indicated by the results of this test?

   _____

   _____

   _____

Problems Encountered

_____

_____

_____

Instructor's Comments

_____

_____

_____

AUTOMATIC TRANSMISSIONS AND TRANSAXLES JOB SHEET 36

Adjusting an External Band

Name _____ Station _____ Date _____

NATEF Correlation

This Job Sheet addresses the following NATEF task:

D.4.5. Inspect bands and drums; determine necessary action.

Objective

Upon completion of this job sheet, you will be able to properly adjust a band with an external adjuster.

Tools and Materials

Basic hand tools

Protective Clothing

Goggles or safety glasses with side shields

Describe the vehicle being worked on:

Year _____ Make _____ Model _____

VIN _____ Engine type and size _____

Transmission type and model _____

PROCEDURE

1. Place the vehicle securely on a lift.   Task Completed ☐

2. Locate the transmission identification tag and identify the exact type and model of transmission you are working on. Where was the ID tag?

   _____

   _____

3. Explain how you know the transmission type from the information on the tag.

   _____

   _____

4. Locate the band adjustment screw. Describe its location.

   _____

   _____

5. Loosen the adjusting screw locknut. If the locknut has a fluid seal, do not re-use the locknut. Install a new nut.   Task Completed ☐

6. Loosen the adjusting screw so that the band can relax around the drum and all tension is off the adjusting screw.   Task Completed ☐

7. Tighten the adjusting screw to the specified torque. What is the torque specification?

   _____

8. Back off the adjusting screw the exact number of turns that are specified in the service manual. How many turns did the manual tell you to back off?

   _____

9. Position a wrench on the adjusting screw and over the locknut so that you can tighten the locknut without moving the adjusting screw.   Task Completed ☐

10. Hold the adjusting screw in position and tighten the locknut to the specified torque.   Task Completed ☐

Problems Encountered

_____

_____

_____

Instructor's Comments

_____

_____

_____